EXPLORING HEALTH AND ENVIRONMENTAL COSTS OF FOOD

WORKSHOP SUMMARY

I0878373

Leslie Pray, Laura Pillsbury, and Maria Oria, *Rapporteurs*

Food and Nutrition Board

Board on Agriculture and Natural Resources

INSTITUTE OF MEDICINE *AND*
NATIONAL RESEARCH COUNCIL
OF THE NATIONAL ACADEMIES

THE NATIONAL ACADEMIES PRESS
Washington, D.C.
www.nap.edu

THE NATIONAL ACADEMIES PRESS 500 Fifth Street, NW Washington, DC 20001

NOTICE: The project that is the subject of this report was approved by the Governing Board of the National Research Council, whose members are drawn from the councils of the National Academy of Sciences, the National Academy of Engineering, and the Institute of Medicine.

This activity was supported by Contract/Grant No. 200-2011-38807, Task Order 4, between the National Academy of Sciences and the Centers for Disease Control and Prevention. The views presented in this publication do not necessarily reflect the views of the organizations or agencies that provided support for the activity.

International Standard Book Number-13: 978-0-309-26580-5
International Standard Book Number-10: 0-309-26580-0

Additional copies of this report are available from the National Academies Press, 500 Fifth Street, NW, Keck 360, Washington, DC 20001; (800) 624-6242 or (202) 334-3313; http://www.nap.edu.

For more information about the Institute of Medicine, visit the IOM home page at: **www.iom.edu.**

Printed in the United States of America

The serpent has been a symbol of long life, healing, and knowledge among almost all cultures and religions since the beginning of recorded history. The serpent adopted as a logotype by the Institute of Medicine is a relief carving from ancient Greece, now held by the Staatliche Museen in Berlin.

Cover credit: Image designed by Casey Weeks.

Suggested citation: IOM (Institute of Medicine) and NRC (National Research Council). 2012. *Exploring health and environmental costs of food: Workshop summary.* Washington, DC: The National Academies Press.

THE NATIONAL ACADEMIES

Advisers to the Nation on Science, Engineering, and Medicine

The **National Academy of Sciences** is a private, nonprofit, self-perpetuating society of distinguished scholars engaged in scientific and engineering research, dedicated to the furtherance of science and technology and to their use for the general welfare. Upon the authority of the charter granted to it by the Congress in 1863, the Academy has a mandate that requires it to advise the federal government on scientific and technical matters. Dr. Ralph J. Cicerone is president of the National Academy of Sciences.

The **National Academy of Engineering** was established in 1964, under the charter of the National Academy of Sciences, as a parallel organization of outstanding engineers. It is autonomous in its administration and in the selection of its members, sharing with the National Academy of Sciences the responsibility for advising the federal government. The National Academy of Engineering also sponsors engineering programs aimed at meeting national needs, encourages education and research, and recognizes the superior achievements of engineers. Dr. Charles M. Vest is president of the National Academy of Engineering.

The **Institute of Medicine** was established in 1970 by the National Academy of Sciences to secure the services of eminent members of appropriate professions in the examination of policy matters pertaining to the health of the public. The Institute acts under the responsibility given to the National Academy of Sciences by its congressional charter to be an adviser to the federal government and, upon its own initiative, to identify issues of medical care, research, and education. Dr. Harvey V. Fineberg is president of the Institute of Medicine.

The **National Research Council** was organized by the National Academy of Sciences in 1916 to associate the broad community of science and technology with the Academy's purposes of furthering knowledge and advising the federal government. Functioning in accordance with general policies determined by the Academy, the Council has become the principal operating agency of both the National Academy of Sciences and the National Academy of Engineering in providing services to the government, the public, and the scientific and engineering communities. The Council is administered jointly by both Academies and the Institute of Medicine. Dr. Ralph J. Cicerone and Dr. Charles M. Vest are chair and vice chair, respectively, of the National Research Council.

www.national-academies.org

PLANNING COMMITTEE ON EXPLORING THE TRUE COSTS OF FOOD: A WORKSHOP[1]

HELEN H. JENSEN (*Chair*), Professor, Department of Economics, Center for Agricultural & Rural Development, Iowa State University, Ames

BILLY COOK, Senior Vice President and Director, Agricultural Division, The Samuel Roberts Noble Foundation, Ardmore, OK

JUSTIN D. DERNER, Rangeland Scientist, Agricultural Research Service, U.S. Department of Agriculture, High Plains Grasslands Research Station, Cheyenne, WY

GREGORY A. KEOLEIAN, Professor and Director, Center for Sustainable Systems, University of Michigan, Ann Arbor

CATHERINE L. KLING, Professor, Department of Economics, Iowa State University, Ames

ROBERT S. LAWRENCE, Professor, Environmental Health Sciences, Health Policy, and International Health, and Director, Center for a Livable Future, Bloomberg School of Public Health, Johns Hopkins University, Baltimore, MD

AARON WERNHAM, Director, Health Impact Project, The Pew Charitable Trusts, Washington, DC

WALTER C. WILLETT, Professor of Epidemiology and Nutrition, Harvard School of Public Health, Boston, MA

IOM and NRC Staff

MARIA ORIA, Study Director
LAURA PILLSBRY, Program Officer
ALLISON BERGER, Senior Program Assistant
EMILY TOMAYKO, Mirzayan Science & Technology Policy Fellow
ANTON L. BANDY, Financial Officer
GERALDINE KENNEDO, Administrative Assistant
LINDA D. MEYERS, Director, Food and Nutrition Board
ROBIN SCHOEN, Director, Board on Agriculture and Natural Resources

[1] Institute of Medicine planning committees are solely responsible for organizing the workshop, identifying topics, and choosing speakers. The responsibility for the published workshop summary rests with the workshop rapporteurs and the institution.

Reviewers

This workshop summary has been reviewed in draft form by individuals chosen for their diverse perspectives and technical expertise, in accordance with procedures approved by the National Research Council's Report Review Committee. The purpose of this independent review is to provide candid and critical comments that will assist the institution in making its published workshop summary as sound as possible and to ensure that the workshop summary meets institutional standards for objectivity, evidence, and responsiveness to the study charge. The review comments and draft manuscript remain confidential to protect the integrity of the process. We wish to thank the following individuals for their review of this workshop summary:

John Blanton, Agricultural Research Programs Manager, The Samuel Roberts Noble Foundation, Ardmore, OK
James K. Hammitt, Professor of Economics and Decision Sciences, Center for Risk Analysis, Harvard University, Boston, MA
Molly Jahn, Professor, Department of Agronomy, University of Wisconsin–Madison
Shiriki K. Kumanyika, Professor of Epidemiology and Associate Dean for Health Promotion and Disease Prevention, University of Pennsylvania School of Medicine, Philadelphia
Stephanie Mercier, Agricultural Policy Consultant, Alexandria, VA
Liz Wagstrom, Chief Veterinarian, National Pork Producers Council, Washington, DC

Although the reviewers listed above have provided many constructive comments and suggestions, they did not see the final draft of the workshop summary before its release. The review of this workshop summary was overseen by **John W. Erdman, Jr.,** University of Illinois at Urbana-Champaign. Appointed by the Institute of Medicine, he was responsible for making certain that an independent examination of this workshop summary was carried out in accordance with institutional procedures and that all review comments were carefully considered. Responsibility for the final content of this workshop summary rests entirely with the authors and the institution.

Contents

1

Introduction[1]

The U.S. food system provides many benefits, not the least of which is a safe, nutritious, and consistent food supply. However, the same system also creates significant environmental, public health, and other costs that generally are not recognized and not accounted for in the retail price of food. These include greenhouse gas (GHG) emissions (Gonzalez et al., 2011); soil erosion, air pollution, and other environmental consequences (Heller and Keoleian, 2003; Wolf et al., 2011); the transfer of antibiotic resistance from food animals to humans (Hayes et al., 2011); and other human health outcomes, including foodborne illnesses and chronic disease (Heller and Keoleian, 2003). Some of these external costs (i.e., external to the food system), which are also known as externalities, are accounted for ("internalized") in ways that do not involve increasing the price of food (see Box 1-1). But many are not. They are borne involuntarily by society at large (Tegtmeier and Duffy, 2004). A better understanding of external costs would help decision makers at all stages of the life cycle to expand the benefits of the U.S. food system even further. The Institute of Medicine (IOM) and National Research Council (NRC), with support from the U.S. Centers for Disease Control and Prevention (CDC), convened

[1] This workshop was organized by an independent planning committee whose role was limited to designing the workshop program and identifying goals, topics, and speakers. This workshop summary has been prepared by the rapporteurs as a factual summary of the presentations and discussions that took place at the workshop. Statements, recommendations, and opinions expressed are those of individual presenters and participants and are not necessarily endorsed or verified by the Institute of Medicine or the National Academies; they should not be construed as reflecting any group consensus.

BOX 1-1
Externality as Defined by Individual Speakers

Katherine Smith defined externality as

> a cost or benefit not transmitted through prices that is incurred by a party who did not agree to the action causing the cost or benefit.

James Hammitt referred to the definition of externality laid out in the National Research Council (2010, p. 29) report *The Hidden Costs of Energy: Unpriced Consequences of Energy Production and Use*:

> An externality, which can be positive or negative, is an activity of one agent (for example, an individual or an organization, such as a company) that affects the well-being of another agent and occurs outside the market mechanism.

a public workshop on April 23-24, 2012, to explore the external costs of food, methodologies for quantifying those costs, and the limitations of the methodologies.

The workshop was intended to be an information-gathering activity only. Given the complexity of the issues and the broad areas of expertise involved, workshop presentations and discussions represent only a small portion of the current knowledge and are by no means comprehensive. The focus was on the environmental and health impacts of food, using externalities as a basis for discussion and animal products as a case study (i.e., specifically beef, poultry, pork, and dairy). The intention was not to quantify costs or benefits, rather to lay the groundwork for doing so. A major goal of the workshop was to identify information sources and methodologies required to recognize and estimate the costs and benefits of environmental and public health consequences associated with the U.S. food system (see Box 1-2). It was anticipated that the workshop would provide the basis for a follow-up consensus study of the subject and that a central task of the consensus study will be to develop a framework for a full-scale accounting of the environmental and public health effects for all food products of the U.S. food system.

Nor was the intention to make any recommendations or suggest policies. Rather, again, it was to lay the groundwork for future efforts. According to Anne Haddix, senior policy advisor at CDC's National Center for Disease Prevention and Health Promotion, the hope is that a framework can be built that will help to identify novel strategies for dealing with food system-related public health problems, such as obesity, in ways that are not only healthful, but also environmentally sound and economically produc-

BOX 1-2
Statement of Task

An ad hoc committee will organize a 1.5-day public information-gathering workshop to examine the challenges inherent in estimating the costs of the U.S. food system not reflected in retail prices and to consider the kind of research strategy that would be needed to approach such a full-scale accounting. The workshop will identify the types of information sources and methodologies required to recognize and estimate the costs and benefits of externalities and unintended public health consequences associated with the U.S. food system.

While the central focus of the workshop will be to understand how to account for externalities and unintended public health consequences of the U.S. food system broadly, meat will be used as a case study with which to explore how to approach the measurement of environmental and public health effects. The workshop planning committee will select the animal species (e.g., beef, pork, chicken, or fish) and different production, marketing, distribution, and retail systems that would provide the most appropriate points for analysis. It is anticipated that the workshop will identify key categories of externalities and unintended public health consequences associated with the production and consumption of meat, the extent of information available on each of the categories, appropriate metrics for quantification, limitations and knowledge gaps, as well as modeling and other analytical approaches needed to establish the value of these costs and benefits.

The workshop would also provide the basis for a follow-on planning discussion involving members of the IOM Food and Nutrition Board and NRC Board on Agriculture and Natural Resources and others to develop the scope and areas of expertise needed for a larger-scale, consensus study of the subject. Based on the framing of the study by the planning discussion, it is envisioned that a central task of the work of a subsequent consensus study committee will be to develop a framework for true-cost accounting of the U.S. food system and to attempt to draw supportable conclusions about the true costs of food.

tive. Currently, no framework is available for analyzing in a comprehensive and systematic way how the food system impacts public health. Although the CDC's initial intention was to focus on public health, Haddix described the food system as being so complex and interactive that it is impossible to separate the health consequences of the food system from environmental, economic, social justice, and other consequences. Thus, the workshop planning committee invited a diverse group of experts and stakeholders to participate in the discussion, including economists, farmers, environmental and agricultural scientists, and public health experts. Their expertise spanned the entire course of the food life cycle.

Given the diversity of perspectives, numerous challenges and complexities regarding the types of information sources and methodologies available to measure the health and environmental costs and benefits associated with the U.S. food system were identified over the course of the workshop. Some participants questioned the rationale for conducting a full-scale accounting of the costs of food and whether another approach might be more feasible. They also stressed that all costs are relative because all food and agricultural systems are dependent on the natural environment; therefore, such an exercise would need to undertake comparisons of alternative food system activities or practices. The heterogeneity of landscapes and management practices among sites only complicates this endeavor, as emphasized by many workshop participants. Participants also expressed varying opinions about the limitations of framing the analysis in terms of externalities. Several other issues were noted, including the broad range of external costs and benefits that were not included in the focus of the workshop; the lack of sufficient data; the importance of considering all stages of the food life cycle; the risks associated with simplifying assumptions about the effects; the inability of models to capture the heterogeneity among food production methods; the variability in the degree of certainty around the magnitude of some effects; and the numerous unanswered questions about the methodologies discussed for quantifying health, environmental, and other effects. Many of these overarching issues are discussed in greater detail in Chapter 7.

By bringing together a wide range of experts, however, the workshop was able to forge connections across subjects that typically are discussed as though they are distinct from one another. The diversity of perspectives and experiences represented among the participants allowed for this workshop to become an important first step in illuminating the range of expertise, methodologies, and information sources that would need to be included in future explorations of the topic.

ORGANIZATION OF THIS REPORT

The organization of this report roughly parallels the organization of the workshop itself (see the agenda in Appendix A). Chapter 2 addresses the economics of food prices and considerations for valuing food. Chapter 3 summarizes the Session 1 presentations on measures and strategies for estimating the external environmental and health impacts of food. Speakers considered the opportunities and limitations of several methodologies: life cycle analysis (LCA), health impact assessment, cost-benefit analysis, multidimensional impact assessment and modeling, and risk assessment. Although the focus of the workshop was on environmental and health costs, a panel session on the social and ecological dimensions of the food supply was held to explore some of the broader impacts. Speakers discussed ecosystem services and disservices, health inequalities, accessibility to food, and animal welfare. Chapter 4 summarizes that panel session. Chapter 5 summarizes the two presentations that focused on methodologies and limitations of attaching monetary value to costs and benefits.

Chapters 6 and 7 summarize group discussion that occurred throughout the course of the workshop, including discussion that occurred during the small working group portion of the workshop. About one-third of the workshop time was spent in small working groups. There were four working groups: energy usage and GHG emissions; soil, water, and other environmental consequences; consequences of antimicrobial use in agriculture; and other public health consequences. The groups were asked to identify effects, methodologies for measuring those effects, and limitations of the methodologies. Chapter 6 includes a summary of these working group discussions. Chapter 7 provides an overview of the major overarching themes from all the open discussions that occurred throughout the workshop, including participants' reflections on key considerations for moving forward with future work in this area.

This workshop summary was prepared by the rapporteurs as a factual summary of the presentations and discussions that took place during the

BOX 1-3
Key Terms Used in This Report

End-of-life: In the context of LCA, end-of-life refers to the stage of the product after preparation and consumption by the conumer. At this stage, the food product is disposed of in some manner (e.g., recycled or placed in a landfill).

Health impact assessment (HIA): HIA is not a single method, but rather a systematic process that uses a wide array of data sources, analytical methods, and stakeholder input to determine the potential effects of a proposed policy, plan, program, or project on the health of a population and the distribution of those effects within the population.

Life cycle assessment (LCA): In the context of the food system, LCA is a tool for examining the environmental impact of a product that covers the impacts of manufacturing, of the upstream production chain (e.g., material extraction, fuels, transportation, etc.) and downstream disposal (e.g., recycling, landfilling, etc.). According to Heller and Keoleian (2003), "a product life cycle approach provides a useful framework for studying the links between societal needs, the natural and economic processes involved in meeting these needs, and the associated environmental consequences."

Life cycle stages: For a food product, the following life cycle stages are considered in the context of economic, social, and environmental sustainability indicators: the origin of the product; agricultural and production conditions; processing, packaging, and distribution of the product; preparation and consumption by the consumer; and the end-of-life of the product (Heller and Keoleian, 2003).

workshop. Neither the workshop nor this summary were intended to be exhaustive explorations of the subject. None of the material summarized here should be construed as reflecting group consensus. For an explanation of key terms used throughout this workshop summary, please refer to Box 1-3.

REFERENCES

Gonzalez, A. D., B. Frostell, and A. Carlosson-Kanyama. 2011. Protein efficiency per unit energy and per unit greenhouse gas emissions: Potential contribution of diet choices to climate change mitigation. *Food Policy* 36:562-570.

Hayes, D. J., H. H. Jensen, L. Backstrom, and J. Fabiosa. 2001. Economic impact of a ban on the use of over the counter antibiotics in U.S. swine ratios. *International Food and Agribusiness Management Review* 4:81-97.

Heller, M., and G. Keoleian. 2003. Assessing the sustainability of the U.S. food system: A life cycle perspective. *Agricultural Systems* 76:1007-1041.

Tegtmeier, E. M., and M. D. Duffy. 2004. External costs of agricultural production in the United States. *International Journal of Agricultural Sustainability* 2(1):1-20.

Wolf, O., I. Perez-Dominguez, J. M. Rueda-Cantuche, A. Tukker, R. Kleijn, et al. 2011. Do healthy diets in Europe matter to the environment? A quantitative analysis. *Journal of Policy Making* 33:8-28.

2

The Economics of Food Prices[1]

Americans spend about 9 percent of after-tax disposable personal income on food—less than anywhere else in the world. That is an "extraordinarily small amount," in Katherine Smith's opinion. Smith, vice president of programs and chief economist at the American Farmland Trust, provided some background on how U.S. market prices of foods are determined and an overview of the types of costs not included in market prices. She also offered some reminders when delving into a discussion on external costs because the average external costs mask a tremendous amount of variation in production methods and geography.

DETERMINING THE MARKET PRICE OF FOOD

What U.S. consumers spend on food is tracked through the Consumer Price Index (CPI), which measures changes in the price level of consumer goods and services purchased by households. Food usually makes up about one-seventh of the CPI (15.256 percent in December 2011), with a substantial change in food prices making a significant difference in the overall CPI. Because food prices are so volatile, with fluctuating food prices masking long-term trends in CPI, food is often removed from the CPI so that longer trends can be observed. The same is true of energy. For example, between January 2009 and January 2012, total CPI fluctuated considerably, falling and rising over time. But when food and energy prices are removed, those

[1] This section summarizes the keynote presentation by Katherine Smith and the discussion following her presentation. For more information, please consider USDA/ERS, 2011a,b,c.

fluctuations disappear, and CPI appears relatively stable over the same time period.

Food prices were forecasted to increase by about 3.5-4.5 percent in 2012, which Smith said is within the high end of the normal range.[2] Most of that increase was among animal products (meats, poultry, fish, and fats and oils), as well as cereals/bakery and fruits/vegetables; most of it was a result of increasing commodity and energy prices. Commodities serve as the basis for production, which itself accounts for about 10 percent of the U.S. "food dollar" (i.e., a breakdown of expenditures for a dollar of food) (Figure 2-1). Energy accounts for another 5 percent. Together, changes in production and energy costs "really move" food prices from year to year.

Commodity prices themselves are affected by many variables, including energy prices, stocks-to-use ratio, and weather. For example, between 2000 and 2011, fluctuations in the index price of food more or less paralleled fluctuations in the index price for crude oil. With respect to stocks-to-use ratio, which is a measure of how much of a particular commodity is stored in comparison to how much is being used, the lower the ratio, the less stable the prices of food and the more likely that a widespread unexpected event, such as a flood or a drought, impacts those prices. The stocks-to-use ratio for total world grain and oilseeds has been dropping every month since June 2010, exacerbating other factors affecting prices. Finally, weather is another major driver of commodity price fluctuation. For example, a 2010 drought in Russia damaged about 25 percent of the global supply of wheat, driving wheat prices up. Additionally, some longer term trends could affect commodity prices in the future. Prime among these factors are climate change and water scarcity.

In addition to commodity and energy prices impacting short-term fluctuations in food prices, there are other underlying factors impacting long-term trends in food prices. A key one is that the global population is not only growing, but it is also becoming more affluent, driving a greater demand for high-value foods, like meats. When demand goes up, prices go up. A second key trend is the growing demand for biofuels. In the United States, an increasingly larger proportion of the corn crop is being used to produce ethanol. In other countries, an increasingly larger proportion of sugar cane crops is being used for the same reasons. Ethanol production was expected to increase by 333 percent between 2005 and 2030. Likewise, production of biodiesel is rapidly increasing worldwide. In the United States, most biodiesel production uses soybean oil. Again, as demand goes up (i.e., the demand for corn, soybeans, etc.), so too does price (i.e., of corn, soybeans,

[2] This number reflects the forecasted price inflation at the time of the workshop. Given the extreme drought that occurred between the time of the workshop and the publication of this report, a much higher increase in food prices is expected.

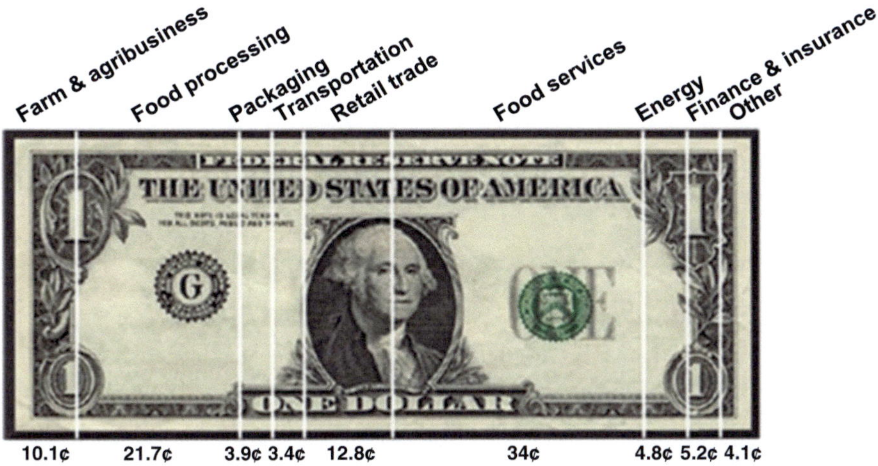

FIGURE 2-1 The industry group food dollar series illustrates the distribution of the food dollar among 10 food supply chain industry groups, demonstrating that the cost of food equals the sum of value added by all supply chain industry groups. NOTE: "Other" includes two industry groups: advertising and legal & accounting. SOURCE: USDA/ERS, 2010.

and other crops). Also underlying these longer term trends in food prices are stagnating investments in agricultural research and development, which slow growth in agricultural productivity.

THE CONCEPT OF EXTERNALITIES: COSTS AND BENEFITS NOT REFLECTED IN MARKET PRICES

Costs not included in the market price of food are called external costs, or externalities. The technical definition of an externality, according to Smith, is "a cost or benefit not transmitted through prices that is incurred by a party who did not agree to the action causing the cost or benefit." General types of externalities associated with food include ecological effects, environmental quality, GHG emissions, animal welfare, social costs associated with labor, and public health effects.

These costs can occur anywhere in the food life cycle, from animal feed crop production through food waste disposal. For an example at the input stage, Smith mentioned potential ecological costs to using genetically engineered crop seeds for animal feed. They include the evolution of newer, more vigorous pests; harm to nontarget species (e.g., butterflies); disruption of biotic communities; and irreparable loss of species diversity or genetic diversity within a species (Snow et al., 2005). The actual crop and animal production stages yield a variety of potential external environmental costs. These can include soil erosion and sedimentation; impaired water quality; overdrawn aquifers; loss of biodiversity; air pollution; and GHG emissions. In some cases, the production stage can include potential external health costs, including pesticide exposure, exposure to pathogens in water systems, inhalation of dust, consumption of chemical residues (e.g., antibiotics and growth hormones), and exposure to antibiotic resistant bacteria. According to Smith, however, this pales in comparison with the direct and indirect health costs associated with food consumption patterns later in the food life cycle.

Even though they have no market value, monetary estimates can be assigned for some of these costs. For example, sedimentation could be valued based on the cost of removing sedimentation from reservoirs and waterways. Or, the cost of pollination services could be used as a proxy for the cost of a loss of biodiversity. Some health costs can be assigned monetary value based on loss of lifespan, loss of work, or related measures.

THINGS TO KEEP IN MIND ABOUT THE EXTERNAL COSTS OF FOOD

Smith offered four "cautions" to the workshop audience—thoughts to keep in mind when discussing externalities of the food system:

1. *Everything is relative.* Costs and benefits are relative. For example, one could argue that agriculture provides a benefit by filtering rainfall and cleaning water. But in comparison to what? That benefit is in comparison to development, not in comparison to a natural ecosystem. The same is true of costs. A cost is only a cost relative to something else.
2. *The average external cost masks variability among producers, consumers, and geography.* For example, one farmer could be doing everything possible to minimize soil erosion, water contamination, and other environmental consequences of production, while his or her "bad actor" neighbor is doing the opposite. Geography plays a role as well. For example, Smith remarked that animal production in dry climates may result in fewer externalities than animal production next to the Chesapeake Bay.
3. *While not in the price of food, we could be paying through the price of other things.* For example, while we may not be paying for sedimentation in the cost of food, we may be paying dredging costs through our water bill. "They can show up in other places," Smith said.

4. *Food and agriculture also have positive externalities.* While the focus of the workshop was on negative externalities, that is, costs, Smith thought it was important to keep in mind that food and agriculture also yield many benefits that are not reflected in the market price of food.

QUESTIONS

Smith fielded several questions on externalities. An audience member asked whether some of the externalities that she mentioned, such as the health costs associated with pesticide exposure, are in fact reflected in the price of food given that organic foods, for example, have higher prices. Smith explained that the higher price for organic foods typically reflects the cost to producers (e.g., the higher cost of allowable practices to control pests).

There was a question about whether the costs of diet-related diseases, such as obesity, can be considered externalities. Smith responded that while those costs are external costs to food, society might be paying for them through health insurance premiums, personal expenses for health and well-being, or other means. An audience member added that the Danish "fat tax" is one way to internalize the external cost of obesity. Smith agreed that taxes can be used to internalize external costs, but cautioned that a food tax typically has to be very high in order to change consumer behavior enough to affect the targeted health outcome.

Another question was about how energy costs should be factored into an analysis of the cost of food given that the actual cost of energy in the United States does not reflect the "true" cost of energy. Smith replied that the actual price of energy used during food production is in fact reflected in the cost of that food. While the discrepancy between the actual and true cost of energy may represent an externality, whether to include that externality in an analysis of the cost of food depends on how far back one wants to go in the life cycle and how many intersecting life cycles one wants to consider. Smith's response led into some further discussion about where to draw the line when thinking about externalities. For example, another audience member wondered whether the profits that other people earn from obesity (e.g., profits from medications prescribed for obesity-related conditions) would be accounted for in an analysis of the external costs of food. Smith replied that it is more manageable to analyze only direct externalities. She said, "Because everything is related to everything else, there are indirect externalities. You can go as far as you want. . . . It gets harder and harder the farther out you go."

REFERENCES

Snow, A. A., D. A. Andow, P. Gepts, E. M. Hallerman, A. Power, J. M. Tiedje, and L. L. Wolfenbarger. 2005. Genetically engineered organisms and the environment: Current status and recommendations. *Ecological Applications* 15:377-404.

USDA/ERS (U.S. Department of Agriculture/Economic Research Service). 2010. *Food Dollar Series*. http://www.ers.usda.gov/data-products/food-dollar-series/food-dollar-application.aspx (accessed September 10, 2012).

———. 2011a. *A revised and expanded food dollar series: A better understanding of our food costs*, ERR-114. Washington, DC: USDA.

———. 2011b. *Have food commodity prices risen again?*, WRS 1103. Washington, DC: USDA.

———. 2011c. *Impacts of higher energy prices on agriculture and rural economies*, ERR-123. Washington, DC: USDA.

3

Understanding Measures and Strategies

Researchers have been examining the health and environmental consequences of the various stages of the food cycle from varying perspectives and with different methodologies. Speakers from several fields were invited to share their approaches to analyzing and interpreting the food system and its unpriced costs (and benefits). This chapter summarizes those presentations.

Marty Heller, research specialist with the Center for Sustainable Systems at the University of Michigan, described how life cycle assessment (LCA) is used to evaluate the environmental impacts of a product. He described the three stages of LCA; discussed how LCA is used, emphasizing that ultimately it is a decision-supporting tool; and considered the opportunities and challenges of using LCA in a study on the cost of food. LCA's greatest strength is its comprehensive nature. It provides a systematic means for analyzing all stages of the food cycle and avoiding "burden-shifting" (i.e., shifting burdens to other life cycle stages, outcomes, or geographic regions). Although Heller did not elaborate, he mentioned the availability of tools that can be used to link LCA results with costing perspectives.

Jonathan Fielding, director of the Los Angeles (LA) County Department of Public Health, described how health impact assessment (HIA) is used to evaluate the health impacts of a policy, plan, program, or project. He emphasized that there is no single HIA approach, rather a range of approaches, and he provided some examples. One of the greatest strengths of HIA, in Fielding's opinion, is the opportunity it provides for intersectoral collaboration during policy decision making and for influencing decision makers to base their decisions on a broader understanding of health and a wider range of evidence.

John Antle, professor in the Department of Agricultural and Resource Economics at Oregon State University, mentioned two additional methodologies that could be used to analyze environmental externalities of the food system: cost-benefit analysis and multidimensional impact assessment. He elaborated on multidimensional impact assessment, emphasizing its reliance on modeling. He also elaborated on some of the major challenges to relying on modeling as a means to quantify externalities. Key among them is the vast heterogeneity that exists in the food system, especially with respect to production (e.g., large- versus small-scale production) and geography, and implications of that heterogeneity for collecting and analyzing data.

Finally, James Hammitt, professor of economics and decision sciences at the Harvard School of Public Health, described how risk assessment could be used to analyze the health outcomes of exposure to a wide range of food system–related stressors; identified sources of data for analysis; and explained how health effects are valued and quantified. Hammitt also discussed, more broadly, the challenge of measuring externalities in the context of noneconomic behavior. He explained that the concept of externality is not very well defined outside the classical economic model. According to classical economic theory, individuals behave as fully informed rational agents. In the "real world," nonmarket factors influence how people behave.

LIFE CYCLE ASSESSMENT[1]

"Eating is an agricultural act."
—Wendell Berry

LCA is a tool for examining the environmental impact of a product. Marty Heller remarked that the defining characteristic of LCA is its "cradle-to-grave" perspective. LCA covers not just the impacts of manufacturing, but also the impacts of the upstream production chain (e.g., material extraction [i.e., mining], fuels and transportation, etc.) and downstream disposal (e.g., recycling, landfilling, etc.). Heller provided an overview of the history of the LCA methodology, described the three main stages of a typical LCA, and discussed how LCA is used and could be used to study the food system.

History of LCA

Heller described the current state of LCA methodology as being in a "mid- to late adolescent stage." The first LCA studies were conducted in the late 1960s and early 1970s on the impacts of different beverage containers, initially for Coca-Cola and later for the U.S. Environmental

[1] This section summarizes the presentation of Marty Heller.

Protection Agency (EPA). Those studies yielded mixed results. Another early series of LCA studies was conducted on the impacts of cloth versus disposable diapers, again yielding mixed results. According to Heller, the mixed results from these early studies were partly a reflection of the variable methodologies being used. At that time, investigators were only just beginning to explore LCA. There was no common theoretical framework upon which to build. The field experienced slow growth in the 1980s, but it did not really "jump forward" into something that "everyone could grab onto" until the 1990s. Since then, the methodology has experienced very rapid growth, with a number of organizations helping to coordinate the harmonization of different theoretical frameworks and to standardize methods and procedures. The EPA, the Society of Environmental Toxicology and Chemistry, and International Organization of Standards (ISO) have all been involved. ISO issued two international standards for LCA, both of which were renewed in 2006: ISO 14040 and ISO 14044. For example, ISO 14040 defines LCA as the "compilation and evaluation of the inputs and outputs and the potential environmental impacts of a product system throughout its life cycle."

Importantly, Heller said, even though ISO has provided good standardization, LCA is still very much an "accommodating methodology." It is used to study a broad range of systems and address a broad range of questions, with many methodological decisions being made along the way. Sometimes, as investigators become familiar with the system under study, decisions made earlier during the LCA may need to be reconsidered. He said, "Appreciating these methodological decisions is pretty important in understanding what the results are really telling us—what we can really draw from those outcomes."

The Three Stages of LCA

Standard LCA has three main stages: (1) goal and scope definition; (2) inventory analysis; and (3) impact assessment. Heller described each in turn.

Goal and Scope Definition

The first stage of LCA involves deciding the purpose of the study, the questions being addressed, and the knowledge being sought. These decisions inform which of two major LCA approaches to take. The typical, or traditional, LCA approach is known as "attributional LCA." Its goal is to describe a system as it is, using data averages. (A "system" includes all the environmentally relevant physical flows in and out of the life cycle and its subsystems.) The second approach is known as "consequential LCA." Its goal is to describe how environmental flows change in response to potential decisions. For example, if a policy decision causes an increased demand for electricity, what are the environmental impacts of that increased demand? Both approaches are important, Heller opined, as they answer different types of questions.

In addition to deciding whether one is going to take an attributional or consequential LCA approach, another important set of decisions made during of the first stage of LCA is defining the system boundaries. Because LCA examines flows between the system being examined and its environment (or another system), it is important to know where that system ends and the environment (or another system) begins. In addition to spatial boundaries, temporal boundaries need to be defined, according to Heller. For example, is the goal to measure impact over the course of a year or over 5 years? For well-defined technical systems, the boundaries can be fairly straightforward. But for agricultural and other biological systems, a distinction between a system and nature (or another system) can be unclear.

Yet another important task of the "goal and scope definition" stage of LCA is defining the functional unit. Heller described LCA as a relative tool. That is, the goal is not to examine absolute impacts, rather impacts relative to some defined unit. The functional unit not only helps to define how flows across a life cycle relate to each other, but it also allows for apple-to-apple comparisons across different systems that produce the same function. Again, with a well-defined technical system, the functional unit is fairly straightforward. For example, the functional unit of an electrical system is the kilowatt hour, with a kilowatt hour generated from a solar panel being relatively the same as a kilowatt hour generated from a coal plant. But it is not clear what the "true" function of food is. It is much more difficult to make that type of direct comparison. Heller explained that many LCAs on food systems have used a reference flow as the functional unit, for example, product mass or volume. While that may be sufficient for benchmarking a product, it does not allow for comparisons across different types of food products. For example, it is difficult to compare a kilogram of beef with a kilogram of milk. Some researchers have explored ways to incorporate nutritional value into the functional unit, for example, grams of protein or caloric value. Foods also have emotional value, although it is unclear how to incorporate that into the functional unit. Depending on stakeholder perspective, one might also think of food as having economic function or environmental function.

Inventory Analysis

The second stage of LCA, inventory analysis, is where Heller said one really "digs in" and examines all of the relevant material and energy flows.

Again, two different main approaches have evolved. First is the traditional "process LCA," which involves examining a very specific product. Heller described process LCA as a "slow" methodology, one that is very data-intensive and typically involves examining very specific processes, yielding very detailed information about a particular product. The second approach, input-output LCA (IO-LCA), uses economic input and output data, usually country-level economic data, and involves examining economic flow between sectors and then connecting those flows with environmental impacts. So rather than looking at a particular type of meat, for example, IO-LCA looks at all meat products. One of the limitations of IO-LCA is that economic input and output data are not always available for all life stages, so some information is missing. Additionally, the level of detail is coarser than what process LCA yields. An advantage of IO-LCA is that it takes the full economy into consideration, which means that decisions do not have to be made about where system boundaries end. IO-LCA captures all interactions among sectors, regardless of how small the contributions of each interaction are to the impact being examined.

Regardless of which approach is taken, process or IO-LCA, inventory analysis is a very data-intensive stage. Often, hundreds of different flows over many dozens of different types of processes are being tracked. Fortunately, Heller said, there is good software available to help organize and account for all those data flows (e.g., SimaPro by Pre Consultants and GaBi by PE International), along with databases that can be used as proxies for some of the ancillary components of the system life cycle. As just one example, the U.S. Department of Agriculture has begun converting some of its agricultural Census data and other information about agriculture into data that can be used in LCA (e.g., www.lcacommons.gov). When good data are not available, for example, enteric fermentation data from ruminant animals, LCA researchers rely on modeling.

In addition to data considerations, another important component of inventory analysis is allocation. Most processes do not produce single outputs, but rather multiple coproducts. This creates a challenge: How should the various emissions and associated environmental burdens be allocated among coproducts? ISO established a hierarchical procedure for managing allocation issues, with their first suggestion being to avoid allocation whenever possible, either by dividing the system into subsystems or by expanding the system. But that is not always possible. For example, it is difficult to divide the process of producing milk from the process of producing meat. If allocation cannot be avoided, ISO's next best suggestion is to reflect the coproduction in some physical relationship between the products (e.g., through energy flows or mass). If that is not possible, the next best suggestion is to reflect it in some other sort of relationship between the products, for example, through economic value. How allocation is managed can significantly affect the results of an LCA. Heller said, "It is important to understand when we are looking at an LCA study—what were some of the decisions that were made here, and how is that influencing what we are looking at?"

Impact Assessment

The third and final stage of an LCA, impact assessment, involves interpreting the environmental significance of the examined material and energy flows. Commonly assessed impacts include energy use, global warming potential, eutrophication, acidification, and tropospheric ozone. Heller also listed some other relevant impact categories that have not received as much attention by LCA investigators: land use, water use, biodiversity, human toxicity, ecotoxicity, erosion, and landscape quality. Depending on the impact being examined, spatial information may or may not be important; for example, the impacts of water use are very spatially dependent. Most LCAs to date have been conducted without much spatial information.

Two additional tasks one may want to consider during this final stage of LCA are normalization, which involves comparing impacts from the system in question to total impacts in a region, and weighting impacts based on the relative importance to society or a particular stakeholder, which Heller said has no scientific foundation, but could be useful when communicating results of the LCA.

Uses of LCA

Heller listed several general uses for LCA. First, it can be used to identify and evaluate unintended consequences. Second, because of its comprehensive nature, it can be used to identify and avoid burden shifting (i.e., avoid shifting burden to other life cycle stages, environmental impacts, or geographic regions). Third, it can be used to identify hot spots. That is, where in the life cycle are the bulk of impacts occurring, and therefore, where should abatement strategies be focused? Fourth, it can be used to communicate environmental impacts, either to consumers or to other stakeholders. Fifth, it can be used to examine differences in scenarios. Once an LCA model has been built, it is easy to examine influences of changes in a particular scenario on environmental impacts. Finally, LCA is ultimately a decision-making tool, one that provides decision makers with more information. Heller noted that both the European Commission and the EPA have used LCA. The European Commission has used it to analyze the impacts of products; the EPA used LCA to determine whether renewable fuels meet GHG thresholds under the Energy Independence and Security Act of 2007.

With respect to the use of LCA in food and agriculture, Heller observed that an International Conference on LCA in the Agri-Food Sector has been meeting since 1996. The group will be meeting for its eighth conference in October 2012. Its first U.S. meeting will be held in 2014. Additionally, there have been hundreds of LCA studies of food and agriculture products, mostly in Europe, although many commodity groups in the United States are beginning to use LCA as a way to quantify environmental impacts (e.g., the Innovation Center for U.S. Dairy is using LCA to examine fluid milk and cheese). Heller and Keoleian (2003) examined key economic, social, and environmental sustainability indicators at each life cycle stage of the U.S. food system (i.e., indicators demonstrating whether the food system is sustainable). Heller mentioned one result in particular that he said always strikes him: 30 percent of energy input into the U.S. food system is associated with household storage and preparation, mostly refrigeration.

Use of LCA in a Study on the Cost of Food: Opportunities and Challenges

In conclusion, Heller encouraged the use of LCA as a tool for addressing at least some questions about the true cost of food. LCA is a comprehensive methodology with established methods and standards. Moreover, not only does it provide a good systematic means to connect production and consumption, but there are other tools available for connecting LCA results with costing perspectives. LCA results can also be linked to health or other additional endpoints, including what Jolliet et al. (2003) refer to as "damage category" endpoints (see Figure 3-1). So, for example, when examining climate change as part of an LCA, one might be interested in further examining how climate change impacts human health or one of the other damage categories. According to Heller, while linking LCA results to additional endpoints adds more levels of uncertainty, a number of methodologies are available for making those links.

However, the methodology is not without its challenges. Data are certainly a challenge, as is the need for some of the methods to be refined for food and agriculture applications. Also, although the focus of LCA is on environmental impact, the methodology is being expanded to incorporate some social impacts (social-LCA), including rural community vibrancy, farmer/worker rights, and eater health (i.e., the impact of the actual food on health). In Heller's opinion, use of the methodology for studying food systems would also benefit from being expanded to incorporate ecosystem services (see the summary of Scott Swinton's presentation in Chapter 4 for discussion on ecosystem services). Currently, LCA is used mostly to assess the impacts of technical systems on the environment, usually with the goal of producing more of whatever that system produces for less impact. Agricultural systems are more challenging than most of these other technical

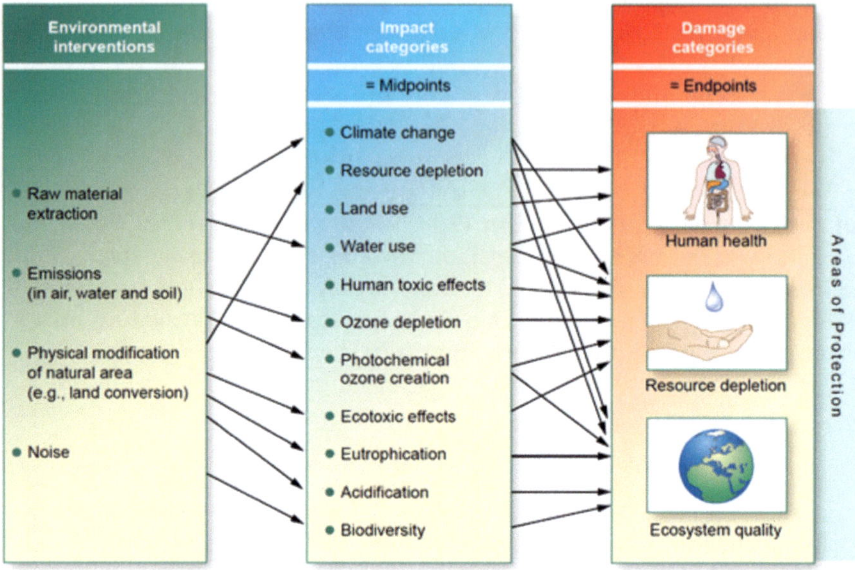

FIGURE 3-1 Overall scheme of the environmental life cycle impact assessment framework linking life cycle impact results to midpoint categories to damage categories.
SOURCE: UNEP/SETAC (adapted from Jolliet et al., 2003).

systems. They are ecosystems—and ecosystems typically have distinct carrying capacities, or the maximum load that can be supported indefinitely by the environment without deterioration; exceeding that carrying capacity can have dire consequences, regardless of whether the system is able to produce more for less impact.

HEALTH IMPACT ASSESSMENT[2]

HIA is not a single method, but rather a systematic process that uses a wide array of data sources, analytical methods, and stakeholder input to determine the potential effects of a proposed policy, plan, program, or project on the health of a population and the distribution of those effects within the population. Additionally, HIA provides recommendations on monitoring and managing those effects. Jonathan Fielding emphasized that

[2] This section summarizes the presentation of Jonathan Fielding.

there is no single HIA approach, rather a wide range of HIA approaches. He discussed the benefits of HIA, described what a typical HIA involves, and provided some examples of national and local HIAs.

The Benefits of HIA: Opportunity for Intersectoral Collaboration

HIA has many benefits. One of its major benefits is that it brings partners from the non-health and health sectors together so that the impact of non-health-sector policy decisions on health is considered during the decision-making process. Fielding observed that many policy decisions in non-health sectors impact public health. For example, agricultural subsidy policy can impact the availability of nutritious foods, mass transportation policy can impact opportunities for walking and bicycling, and environmental policy can impact the availability of clean air. HIA can ensure that health effects are at least on the table for discussion; tip the scales, adding evidence in favor or against a certain course of action; change the terms of debate by encouraging transparent decision making and consideration of the best available evidence; tweak plans, with policies modified in ways that minimize potential harm and maximize potential benefits; bring new parties to the table that give voice to concerns of affected stakeholders who otherwise have difficulty making their concerns known; and change institutional missions and responsibilities. Fielding mentioned the city of Los Angeles as an example of an institution that has incorporated health effects as part of its general plan. Now, when thinking about new developments or policy, potential health effects must be considered. "This seems pretty obvious," Fielding said. "But it was actually a major victory to get health incorporated into the plan."

In addition to its role in facilitating intersector collaboration, HIA also highlights potentially significant health impacts that are unknown, under-recognized, or unexpected; assesses how proposals, policies, and plans will affect all community members, particularly the most vulnerable; supports sustainable development by considering both short- and long-term impacts; and identifies opportunities to enhance health benefits and mitigate harms.

What an HIA Looks Like

There are two kinds of HIA: project-specific and policy-oriented. With project-specific HIA, the major goal is to gather a wider range of stakeholder input into the decision-making process and to make the process more transparent. Policy-oriented HIA affects public policy in a broader way.

Whether the focus is on a specific project or a broader policy, the HIA framework has six major steps:

1. *Screening:* getting a preliminary sense of how important a particular proposal is and whether an HIA would be helpful.
2. *Scoping:* building what Fielding described as the "logic framework," that is, a summary and descriptions of relevant pathways and likely health effects, research questions, and alternatives to the proposed action. For example, if the proposed action is widening of a road, is that really the best approach, or should an increase in mass transit or other actions also be considered as alternatives?
3. *Assessment:* determining the baseline health status, analyzing the beneficial and adverse health effects of the proposed actions and its alternatives, and integrating stakeholder input into the analysis.
4. *Recommendations:* identifying alternatives or actions to avoid or minimize adverse health effects; and proposing a health management plan to identify who could implement those recommendations and, more importantly in Fielding's opinion, monitor implementation of the HIA recommendations.
5. *Reporting:* documenting the HIA and communicating results and recommendations to decision makers, the public, and stakeholders. HIAs come in many forms. An HIA can be a comprehensive 200-page report, a 2-page policy brief, a "logic framework" and supporting discussion showing causal pathways, a checklist completed by an agency or policy makers, a spreadsheet or "calculator" allowing users to estimate health impacts for different scenarios, or a process for guided community engagement that results in testimony given to a policy-making body.
6. *Monitoring and evaluation:* evaluating whether the HIA was conducted according to plan and applicable standards, whether the HIA influenced the decision-making process, and, when practicable, whether implementation actually changed health indicators as expected.

Although the National Environmental Policy Act of 1969 requires an analysis of health effects of proposed actions,[3] Fielding said the requirement was not enforced. Had it been enforced, HIA as a key set of approaches to considering health effects would have probably come into use much earlier. However, the HIA "movement" didn't really start until the mid-1980s in Europe. In the United States, the methodology did not really pick up until around 2000, with only about three dozen U.S. HIAs conducted by the mid-2000 decade. Many more have been conducted since then—both project-specific and policy-oriented HIAs. Most of the growth has been local, with 86 registered HIAs in 2012 being local, 18 state-level, and 6 federal-level.

[3] 42 U.S.C. §§ 4321-4347.

Today, HIA plays an important role in the assessment, monitoring, evaluating, and dissemination component of *Healthy People 2020*.[4]

Examples of HIA

As an example of a national-level, policy-oriented HIA, Fielding described the results of an HIA on the 2002 Farm Bill and its expected health effects on the U.S. population (Partnership for Prevention/UCLA School of Public Health, 2004). Fielding and colleagues identified five major pathways through which the new legislation was expected to impact health: (1) rural income and quality of life; (2) dietary consumption patterns; (3) food safety; (4) environmental pollution; and (5) other environmental degradation. Investigators focused their analysis on two pathways: dietary consumption patterns and environmental pollution. Specifically, they asked whether dietary consumption patterns would be affected by the Farm Bill subsidy policy and whether air pollution would be affected by ethanol production. Results of the HIA indicated, first, that changes in commodity supports would have little, if any, effect on consumer prices and consequently consumption, primarily because commodity price is only a small portion of the consumer price for most foods; second, that the air pollution effects of the production of ethanol were uncertain.

Fielding emphasized the importance of repeating HIA when new information becomes available. When the 2002 Farm Bill analysis was conducted, competing models yielded contradictory results with respect to the air pollution effects of ethanol production. Since then, more recent data have become available that show a negative effect given current production technology.

As an example of a local policy-oriented HIA, Fielding described an LA County HIA of a county government food procurement policy on sodium reduction (Gase et al., 2011).[5] LA County government is large and complex, with over 101,000 employees and 37 departments. The county

[4] Launched in 2010, *Healthy People 2020* is a U.S. government 10-year goal for health promotion and disease prevention. As Fielding explained, the *Healthy People 2020* ecological model of health is an action model, with interventions (i.e., policies, programs, information) impacting not just individual behavior, but also social networks (including family and community networks), living and working conditions, and broader environmental and other conditions (broad social, economic, cultural, health, and environmental conditions and policies at the global, national, state, and local levels). HIA is used to monitor various health outcomes and decide which interventions to support.

[5] Food procurement policy is only one component of the effort to reduce sodium intake. The county is also participating in a national coalition that is encouraging food processors to voluntarily reduce sodium in their foods, and communicating with the U.S. Food and Drug Administration about changing sodium from a generally recognized as safe food to another type of food.

operates many food service venues and contracts with many food service vendors across its 87 regional and local parks, 344 miles of trails, 19 public golf courses, and a multitude of beaches, museums, libraries, theaters, prisons, juvenile prisons, juvenile detention centers, and probation camps. In 2009-2010, the county conducted an HIA to evaluate the likely impact of food procurement policies on sodium intake among patrons at LA County venues. HIA investigators estimated the reach of food procurement policies (e.g., county hospital cafeterias serve about 600 adults per day, senior meals programs serve about 9,200 adults per day, child care venues serve about 24,000 children per day, and other county cafeterias serve about 1,800 adults per day); gathered qualitative data and input from county food service vendors; and mathematically simulated the effects of varying levels of sodium reduction on mean systolic blood pressure (SBP). Preliminary findings predicted that adults eating at county food service venues that reduced sodium levels would consume 233 fewer milligrams of sodium per day, corresponding to a reduction in SBP among adults with hypertension, fewer cases of uncontrolled hypertension, and lower costs of treatment. Thus, the HIA investigators concluded that food procurement policy could have a positive health impact in LA County.

Opportunities and Challenges for Using HIA in a Study on the Cost of Food

In conclusion, Fielding emphasized that HIA is only one of many tools that can be used to inform and improve health policies. But one of its greatest strengths is the opportunity it provides for intersectoral collaboration. He said, "In our experience it has really moved health into discussions in other sectors where we know that there are health impacts of decisions, but health effects have really not been considered."[6] But it is not without its challenges. In many cases, HIA yields only qualitative results or quantitative results with very wide confidence intervals. Its applicability is also challenged by the complexity of relationships between determinants and health outcomes; the lack of research on many causal pathways; the lack of sufficient data on interventions to improve health status; and the lack of reliable and valid indicators of environmental effects.

ENVIRONMENTAL CONSEQUENCES[7]

Measuring the cost of food is a complex, multidisciplinary challenge—one without a magic fix or solution. "We are all groping for how to deal

[6] Fielding referred workshop attendees to the UCLA Health Impact Assessment Clearinghouse Learning and Information Center website: www.HIAguide.org.

[7] This section summarizes the presentation of John Antle.

with [it]," John Antle remarked. Antle shared some thoughts about the challenges of valuing social costs and benefits and discussed the opportunities and challenges of a multidimensional impact assessment modeling approach that he and colleagues have been using to quantify environmental externalities, one that considers not just environmental but also economic and social outcomes.

The Challenge of Valuing Social Costs and Benefits

Traditional agricultural policy is focused on farmer income. Antle speculated on how consideration of the cost of food shifts the focus to a different set of questions that revolve not around transferring income to farmers, rather on social well-being: First, are we producing and consuming the right amounts of food in the right ways? Second, given that many experts would agree that the answer to the first question is no, what can be done to correct the problem(s)?

Antle explained that economists address these questions by thinking in terms of social costs and benefits and attaching value to them. He observed that there is a huge body of empirical research that at least attempts to quantify both negative externalities (i.e., social costs) and positive externalities (i.e., social benefits) associated with agricultural production and food systems. Much of the work to date has focused on farm-level production, as opposed to components of the food system that reside beyond the farm. Despite these efforts, quantifying social benefits and costs in a meaningful way, that is, in a way that can inform public policy, remains tremendously challenging.

Even deciding whether an externality is positive or negative can be challenging. For example, water contamination associated with agriculture could be considered either a positive or negative externality depending on how property rights are initially assigned. One could consider farmers the "bad guys" and tax them accordingly, or as "stewards of the land" and pay them to do more good. In this example, the property rights issue is whether farmers have a right to use fertilizer. If the answer is no, then farmers who pollute the water should be penalized for doing so. But if the answer is yes, that is, if farmers have the right to use fertilizer, then farmers who use less fertilizer should be compensated for the cost they will bear associated with using less fertilizer. A related challenge is understanding how the economic agents that create the externalities, that is, the farmers, respond to policy intervention. According to Antle, modeling that response to policy is another "big part" of what economists do in their effort to answer these questions.

Added to the challenge of quantifying externalities is the reality that, as Antle put it, "There is a lot more to life than externalities." He suggested that food policy may need to address the broader set of factors that people care about, such as food quality or animal welfare.

Additional Tools for Quantifying the Cost of Food

In addition to LCA and HIA, Antle identified two additional tools for quantifying externalities: (1) cost-benefit analysis and (2) multidimensional impact analysis. Additionally, regardless of methodology, he encouraged more appreciation for qualitative considerations and how those could be added to the "quantitative toolbox." Many environmental outcomes, like biodiversity, are difficult to quantify.

Cost-Benefit Analysis

Cost-benefit analysis involves valuing all benefits and costs in monetary terms and then adding those values. The method is limited by many factors, including distributional issues (i.e., the distribution of benefits and costs), the timing of benefits and costs (i.e., comparing current versus future benefits and costs), and valuation. Antle noted that the issue of valuation is especially challenging when considering the cost of food. It is not clear whose values should be used. Another major challenge is aggregation across outcomes, that is, measuring all outcomes in terms of one metric (e.g., measuring all GHG emissions in terms of carbon equivalents). The challenge of aggregation is what led Antle to his work on what he calls "multidimensional impact assessment." Also, Antle questioned the practicality of adding all of the positive and negative externalities associated with food into a single sum, as a cost-benefit analysis would do. He suggested focusing on a small number of key indicators, being careful not to leave out any important ones, and understanding those indicators well enough that their estimated values can actually have a policy impact.

Multidimensional Impact Assessment

Multidimensional impact assessment is a modeling methodology that takes into account economic, environmental, and social impacts—what Antle referred to as the "three pillars of sustainability." The approach involves quantifying key indicators and their relationships, with a focus on trade-offs and synergies. The approach typically involves coupling an ecosystem model with an economic behavioral model and examining and understanding the underlying processes well enough to predict what will happen in response to a policy, price, or other change. Predictions can be made at regional, national, or even global levels. Regional predictions can be made with fairly good site-specific detail. For example, using counties as the basic spatial unit, Antle and Ogle (2012) linked what is known as a century ecosystem simulation model (a model used for examining soil carbon dynamics) to an economic behavioral model and examined the effects of no-till production on GHG emissions, taking into account not just soil carbon, but also nitrous oxide emissions and fuel use.

Data and Modeling Challenges

A major limitation of impact assessment is dimensionality. As Antle explained, trying to quantify trade-offs among 30 key indicators can become extraordinarily difficult. Another major challenge is the heterogeneous nature of agricultural systems. For example, there is considerable heterogeneity even in the corn-soybean world of Iowa or in the wheat-fallow-livestock rangeland systems of the Great Plains. Not only are agricultural systems spatially heterogeneous, but they are also temporally dynamic. Together, this spatial and temporal variation creates very serious analytical challenges to measuring environmental externalities.

In addition to the analytical challenge of quantifying multiple environmental outcomes across space and time, collecting enough high-quality data to conduct those analyses in the first place can also pose a challenge. This is true even though the availability and quality of some types of data have improved over time. For example, farm-level agricultural census and other data are now available (e.g., Agricultural Resource Management Survey data, jointly collected and managed by the Economic Research Service [ERS] and the National Agricultural Statistics Service [NASS]). Also, remotely sensed data are yielding more accurate annual land use and crop yield data, improving the capability to not only conduct multidimensional impact assessment modeling, but also make it more dynamic. Yet, at the same time, federal government budget constraints are reducing availability of other types of data. Antle mentioned that he has not been able to access data more recent than the 2007 version of the National Resources Inventory, which he said used to be a heavily used data source (the database is currently being revised). He said, "If we really want to get at this question of the true cost of food, good data are going to be really essential."

Added to its analytical and data challenges, impact assessment is made difficult by funding and institutional constraints to transdisciplinary team building. It is also costly, Antle cautioned. Indeed, in Antle's opinion, the cost of analysis is "one of the big issues" to consider when planning a study on the cost of food.

Yet another challenge is the lack of a systematic approach to model development, comparison, and improvement—in other words, as Antle put it, to "making this modeling research really good science." He mentioned the Agricultural Model Intercomparison and Improvement Project and its efforts to compare and integrate different modeling approaches.

PUBLIC HEALTH CONSEQUENCES[8]

Based on his work with risk assessment, James Hammitt views health as depending on three sets of factors: (1) exposure to physical, chemical, biological, and social agents and stressors; (2) behavior that reduces exposure or mitigates the effects of exposure, with the ability to avert or mitigate exposure depending on income, information, and other resources; and, more broadly, (3) the food production-distribution-consumption system, which can affect both exposure and behavior. As examples of how the system itself impacts health, specific foods have constituents and contaminants that affect health; overall diet affects health; and the food system itself can impact disposable income, with a higher cost of food reducing the amount of disposable income available for other, health-protective measures. Hammitt discussed how risk assessment can be used to analyze exposures to stressors in the food system and their impact on health, and identified major sources of data for conducting risk assessments. He also offered some general thoughts on defining externality in the context of noneconomic behavior.

Risk Assessment

Risk assessment involves analyzing how exposures to various stressors influence the probability of morbidity and mortality. Risk assessment involves three steps: (1) identify exposure; (2) calculate exposure-response functions; and (3) attach value.

Identify Exposure

Hammitt identified five major exposure pathways in the food system: (1) production and processing (i.e., exposure to nutrients, contaminants, and pesticides in food; waste streams from production and processing facilities; and energy pathways related to production and processing); (2) packaging (i.e., exposure to contaminants in packaging, and to energy pathways related to packaging); (3) distribution (i.e., exposure to energy pathways related to distribution); (4) preparation (i.e., exposure to nutrients and contaminants influenced by how food is prepared and to environmental pollution related to energy use); and (5) consumption (i.e., exposure to nutrients, contaminants, and pesticides in food, and to overall diet).

[8] This section summarizes the presentation of James Hammitt.

Hammitt noted that his list of exposure pathways was "not very well researched" and that his intention was for the list to be "provocative." Another public health effect to consider that he said does not really fit into any one of these five pathways is antibiotic resistance. He suggested that one way to analyze antibiotic resistance is by examining the distribution of resistant microbial strains and the ways that people can come into contact with those strains (i.e., the chance of infection from all possible pathways).

Calculate Exposure-Response Function

After identifying exposure, the next step is to calculate what is known as the exposure-response function, that is, the probability of an adverse health effect given exposure to a certain quantity of stressor.

Attach Value

The third and final step is valuation. Because multiple health effects can arise, with the same food sometimes having both "good" and "bad" effects (e.g., eating fish can be cardio-protective because of its omega-3 fatty acid content, while at the same time serving as a major exposure pathway to methyl mercury, which has negative health effects), valuation involves aggregating those effects not just for individuals, but also across a population (the "social aggregation problem"). Economists attach value in one of two ways: via either monetary value (i.e., willingness to pay [WTP]); or health utility (i.e., quality-adjusted life years [QALYs], value per statistical life, or a related concept).

Analyzing Health Effects

Hammitt offered some thoughts on how one might use risk assessment to analyze health outcomes associated with four major sources of exposure: (1) diet; (2) nutrients, contaminants, and pesticides; (3) energy; and (4) waste streams.

Diet clearly affects the risk of many diseases and health effects, from cardiovascular disease to obesity, not just for the person actually consuming any given diet, but also for his or her offspring. The major source of information for analyzing those effects is epidemiological data. Hammitt commented on how diet itself is affected by many factors, including prices and convenience (i.e., as determined by availability, distribution, and preparation), consumer information about the consequences of eating different types of foods, and traditions and customs. To examine how these many factors impact diet, Hammitt speculated that the major data sources would be economic (e.g., demand system modeling, social science methods that determine how people perceive risk and choose diet).

With respect to exposure to nutrients, contaminants, and pesticides, Hammitt noted that the effects can be either positive or negative and that some substances may have positive or negative effects depending on dose. For example, many nutrients are beneficial at low levels, but harmful at higher levels. Additionally, some contaminants and pesticides may have safe exposure thresholds, below which there is no risk of adverse effect. Major sources of information for analyzing these effects include epidemiology and toxicology data. One of the challenges to collecting and analyzing these types of data is that the dose-response function is often unknown or unreported. While some nutrients have established tolerable upper levels (ULs), above which risk for adverse effects increases, for other nutrients, not enough data were available to establish ULs.

All stages of the food life cycle use commercial energy. Hammitt identified production and processing, packaging, and possibly distribution as the most energy-intensive stages, with the main exposure pathways being environmental release of stressors, mostly air pollutants. The major source of exposure is fossil fuel combustion (i.e., for electricity production, farm vehicle transportation, etc.), with particulate matter and, to a lesser extent, ozone precursors being especially problematic. Even after consumption, waste disposal and clean-up processes use energy (e.g., hot water to wash dishes), although it is unclear whether postconsumption energy use is quantitatively significant. Upstream pollution associated with producing these fuels in the first place is another source of exposure to consider. Hammitt remarked that research on these phenomena is a fairly well-developed field, with major information sources being epidemiology; and fate, transport, and exposure modeling.

Finally, waste streams are another important type of food-related exposure pathway. Waste streams include waste from livestock and fertilizer/pesticide run-off from fields. Waste stream exposure pathways include air emissions; water exposure (i.e., through drinking, bathing, swimming); and food contamination (e.g., livestock waste is the source of many bacterial outbreaks in vegetables). Again, data for studying these types of exposure and their health effects come from epidemiology; toxicology; and fate, transport, and exposure modeling.

Valuing Health Effects

There are two conventional approaches to valuing health effects. The first is a money measure: WTP, or willingness to accept compensation for change. In economic parlance, WTP is the change in wealth that one is willing to accept to increase his or her survival probability by a certain amount. WTP is widely used in environmental and transportation applications. The second approach is based on health utility: QALYs. QALYs are a measure of the trade-off between health (or, more specifically, health-related quality of life, with "0" being death and "1" being optimal health) and longevity. QALYs are estimated using what is known as a health profile, that is, a graph with time on the x-axis and health-related quality of life on the y-axis, and with total QALYs being equal to the area under the curve.[9] QALYs are widely used in public health and medical applications.

A key question to consider when thinking about these two different approaches is whether they are consistent with the preferences of the affected people. Hammitt explained how economists traditionally have considered individuals to be the best judges of their own interests and determined whether a policy is going to improve someone's situation by asking him or her whether they think they would be better off with the policy in place. He said, "It seems like it is very important, if we are trying to measure welfare that the welfare measure ought to at least have something to do with the preferences of the people whose welfare we are trying to affect." A significant difference between the money measure and health utility approaches is that the health utility approach is based on assumptions about individual preferences that are reasonable on average, but clearly false at the individual level. For example, one assumption is that the trade-off between health and length of life that QALYs measure is independent of wealth (i.e., that the fraction of one's lifetime one would give up to be free of some disability is independent of whether one can afford technologies that help offset the disability, such as reading glasses or electric wheelchairs), an assumption that Hammitt said does not make much sense either theoretically or empirically. WTP is less constrained and more accurately reflects individuals' own preferences. So, for example, even if the risk of cancer from smoking were exactly the same as the risk of cancer from pesticides in food, an individual might have different preferences for those two risks and, therefore, differ-

[9] During the question-and-answer period, Hammitt explained that QALYs are estimated by surveying and interviewing people. People are asked three general types of questions: (1) Respondents are asked to rate their health on a scale from 0 to 100. (2) They are asked what is called a "time trade-off" question. For example, they are asked to assume that they are going to live the rest of their life, say 40 years, in a specific health state with a chronic illness and decide at what point they would exchange that future life for a shorter life in perfect health. (3) Respondents are asked what is called a "standard gamble" question. They are asked to imagine living the rest of their life in an impaired health state and being given the option for a treatment that would either restore their health or cause death. Surgery is a good example of this type of treatment, that is, one with an up-front mortality risk. Hammitt said that there is a lot of "noise" associated with each of these three questions, but that they generally provide the same answers.

ent willingness to pay to reduce those risks. WTP would account for those different preferences, while QALY would not. However, to the extent that people are confused about their own preferences, empirical estimates of WTP are more susceptible to fuzzy thinking.

But when is it appropriate to harm some people in order to provide benefits to others? The implicit social objective to increasing QALYs is to maximize total health and longevity in a population. A policy that provides more total QALYs to one subpopulation is viewed as better than a policy that provides fewer total QALYs to another subpopulation, regardless of the number of people in each group and their characteristics. The same situation occurs with WTP. The implicit social objective to making a policy based on WTP is to maximize total WTP, independent of the number and characteristics of the people who benefit.

Defining Externality in the Context of Noneconomic Behavior

The concept of externality is not well defined outside the classical economic model—that is, when individuals do not behave as fully informed rational agents as economic theory assumes they do. Hammitt identified several questions to consider before embarking on a study of the external costs and benefits of food. First, when human behavior differs from what an economic model assumes, how should externality be defined? Will the study really be about externalities, or will it be about consequences? Why focus on externalities? Is it because, if all externalities could be internalized, then the outcome would be efficient? While that may be the case in a simple economic model, it is not true in the real world. In the real world, nonmarket influences affect how people behave. For example, information asymmetry, that is, where people do not know about the properties of various foods or the risks of different diets, influences eating. Another major nonmarket influence is the social network or environment, as people are influenced by what other people in their social networks do. Another question to consider is whether health risks from poor diets are internalized or not. At the simplest, one can assume that consumers are the best judge of their own interests. But again, behavioral economics and related research suggest that people do not behave as simple economic theory assumes they do. People behave inconsistently over time and exhibit limited self-control, for example, by procrastinating (i.e., eating badly today while promising to eat better tomorrow). It is unclear whether the health risks associated with those behaviors constitute externalities.

REFERENCES

Antle, J., and S. Ogle. 2012. Influence of soil C, N_20, and fuel use on GHG mitigation with no-till adoption. *Climatic Change* 111(3):609-625.

Gase, L. N., T. Kuo, D. Dunet, S. M. Schmidt, P. A. Simon, and J. E. Fielding. 2011. Estimating the potential health impact and costs of implementing a local policy for food procurement to reduce the consumption of sodium in the County of Los Angeles. *American Journal of Public Health* 101(8):1501-1507.

Heller, M., and G. Keoleian. 2003. Assessing the sustainability of the U.S. food system: A life cycle perspective. *Agricultural Systems* 76:1007-1041.

Jolliet, O., A. Brent, M. Goedkoop, N. Itsubo, R. Mueller-Wenk, et al. 2003. *Final report of the LCIA definition study.* http://www.lca-net.com/files/LCIA_defStudy_final3c.pdf (accessed October 19, 2012).

4

Examining Social and Ecological Costs and Benefits

Although the main focus of the workshop was on health and environmental costs, Anne Haddix stated in her introductory remarks that examining one outcome in isolation is practically impossible. The food system is inordinately complex, with multiple and intersecting inputs and outputs, costs and benefits, and units of analysis. Several speakers were invited to share their thoughts on some of the broader social and ecological costs and benefits of the food system. This chapter summarizes their presentations.

Scott Swinton, professor and associate chair in the Department of Agricultural, Food, and Resource Economics at Michigan State University, discussed the ecological costs and benefits of food production using an "ecosystem services" concept and framework that was developed as part of a United Nations multiyear study on the consequences of ecosystem change for human well-being. He also explored the challenges of valuing nonmarket ecological costs and benefits. Key among those challenges is that ecological costs and benefits are highly variable across place and time.

Steven Wing, professor of epidemiology at the University of North Carolina at Chapel Hill, urged workshop participants to think not just about cost, but cost to whom. In his opinion, many of the animal production costs discussed during the workshop are related to health inequalities. He reflected on how communities where concentrated animal feeding operations (CAFOs) are located—and the individuals who live in those communities—pay a disproportionate amount of the external cost for animal production in loss of health and quality of life (Donham et al., 2007; Horton et al., 2009; Lipscomb et al., 2005, 2007a,c, 2008; Mirabelli et al.,

2006; Schinasi et al., 2011; Tajik et al., 2008; Wing and Wolf, 2000; Wing et al., 2008). He commented on how animal consumption costs are related to health inequalities as well, with people who live in low-income areas having limited food choices.

Ricardo Salvador, director and senior scientist in the Food and Environment Program at the Union of Concerned Scientists, expanded on the theme of food choice. He discussed dynamics among poverty, food insecurity, and health and made the case that health is partly a reflection of one's environment and that not everyone has access to the same food choices. He argued that accessibility to food is a social issue, not just an economic issue, and therefore that using an economic model as a framework for studying the cost of food limits what can be detected.

Finally, Jayson Lusk, professor and Willard Sparks Endowed Chair in the Department of Agricultural Economics at Oklahoma State University, described the costs and benefits of animal welfare legislation and methods used to value those costs and benefits. He emphasized the importance of trade-offs when analyzing the cost of food. With respect to animal welfare, the question is not the well-being of animals; the question is, what do we have to give up to attain that benefit?

AGRICULTURAL ECOSYSTEM SERVICES AND THE COSTS OF FOOD PRODUCTION[1]

Food production systems can be thought of as agricultural ecosystems that are managed to provide food. In other words, according to a framework laid out in the Millennium Ecosystem Assessment (MA),[2] food production systems yield what are known as "provisioning" ecosystem services. Food production systems also generate a suite of other, nonprovisioning ecosystem services (e.g., some farms provide aesthetic services, others provide fiber and bioenergy); they also rely on various ecosystem services (i.e., services that allow crops to grow, soil to form, etc.) (Swinton et al., 2007). Of course, not all ecosystem inflows and outflows are desirable. Ecosystems also produce costs. For example, food production can

[1] This section summarizes the presentation of Scott M. Swinton.

[2] Called for by the United Nations Secretary-General Kofi Annan in 2000, the MA assessed the consequences of ecosystem change for human well-being and the scientific basis for action needed to enhance the conservation and sustainable use of those systems and their contribution to human well-being. Results of the assessment were published in a series of reports in 2005. The MA defined four basic types of ecosystem services: provisioning (i.e., provision of food, fiber, fuel), regulating (i.e., regulation of climate, water, and habitat), supporting (i.e., support of other ecosystem services through soil formation, nutrient cycling, primary productivity, etc.), and cultural (i.e., aesthetic, recreation, scientific knowledge, and other cultural services). For more information, visit http://www.maweb.org/en/Index.aspx.

negatively impact water (i.e., water pollution), health (i.e., increased health risks from hormones, agrochemicals, antibiotics), climate change (i.e., from greenhouse gas [GHG] emissions), habitat of desirable species (i.e., habitat loss), and aesthetics. Plus, food production itself can be negatively impacted by pests, disease, and other detrimental factors that reduce productivity or increase production costs (also referred to as ecosystem disservices). Swinton discussed methods for valuing ecosystem services and disservices associated with food production systems; and the challenges of nonmarket valuation (i.e., intrinsic value; see Box 4-1).

Nonmarket Valuation of Ecosystem Services and Disservices

Another way to think about a food production system, or an agricultural ecosystem more broadly, is as a transformation process, with both synthetic and ecosystem inputs feeding into the process and food and ecosystem outputs coming out the other end. External costs can accrue on either the input or output side. For example, costs occur on the input side when natural capital is depleted (i.e., the natural capital that is necessary for enabling the provisioning ecosystem services provided by animal production), such as overgrazing of rangeland. Costs occur on the output side when natural capital is contaminated, such as water pollution from animal feeding operations. Because markets are often absent from this transformation process, valuing these costs is challenging.

Attaching dollar values to ecosystem services linked to food production involves, first, measuring changes in quantity (i.e., measuring the baseline production process and then measuring changes associated with each alternative feasible process), and second, associating values with those changes. Estimating those values involves examining both the supply and demand sides. On the demand side, what would people who do not have something be willing to pay in order to get more of it? That is, what is the willingness to pay (WTP) for a change in ecosystem service? On the supply side, what

BOX 4-1
Economic Versus Intrinsic Value

Swinton emphasized that economic values resulting from the relation between supply and demand are not the same as intrinsic values. In his book *Nature and the Marketplace*, Geoffrey Heal (2000) uses water and diamonds to explain the difference. Water, which is essential to human life, has low economic value but high intrinsic value. On the other hand, diamonds, which are not essential to human life, have low intrinsic value but high economic value.

are suppliers willing to accept in order to change something? That is, what is the willingness to accept for a change in a production cost (e.g., reduction in profitability) associated with a modified practice?

Nonmarket valuation methods attempt to simulate supply and demand where markets do not actually exist. Without going into detail, Swinton said, many different nonmarket valuation methodologies are used, most of which are cost-based (cost of remediation, factor substitution, production function, travel cost/cost of illness), although some are based on stated preference (contingent valuation/ranking) (Freeman, 2003; MA, 2003). The methods are based on a core set of principles adapted from Bockstael et al. (2000) and Pearce (1998): (1) marginal changes from a baseline occur within a range that can be observed; (2) budget constraints limit the choices that can be made; and (3) decision makers select the best alternative, even if that alternative is very different from the original choice.

Challenges to Nonmarket Valuation

Nonmarket valuation is complicated by several factors. First, the same ecosystem service can have both external and internal effects, or costs. For example, erosion control can have effects off-farm, such as waterway and reservoir siltation, that occur unwillingly and without the involvement of others. Those are external costs. Erosion control can also have an internal cost if the farmer makes an erosion control management decision aimed at improving crop yield. Second, ecosystem services are experienced at varying scales. For example, erosion control off-farm is experienced at a watershed scale, whereas erosion control on-farm is experience at the farm scale. As another example, climate regulation services are experienced at a global scale, but pollination and genetic diversity services (e.g., pest control) are experienced at local or regional levels. Third, ecosystem services are often bundled together on the supply side through a production system, yet consumers experience those services individually. For example, consumers may experience improved drinking water (an individual event) that occurred as a combination of several agricultural practices and environmental changes (see Figure 4-1). So there is often a big gulf between valuation on the production side and what people experience on the consumer side. Changing one ecosystem service may require changing the entire production process, causing a whole set of intermediate environmental changes (e.g., reducing nitrogen fertilizer use in order to cut GHG emissions from nitrous oxide also reduces nitrate leaching). The consumers experience those changes differently (e.g., they experience less fertilizer input as improved drinking water quality) (Chen, 2010). Fourth, the value of ecosystem services varies across time and space, because supply and demand vary across time and space. This variation makes it very difficult to estimate values for widely

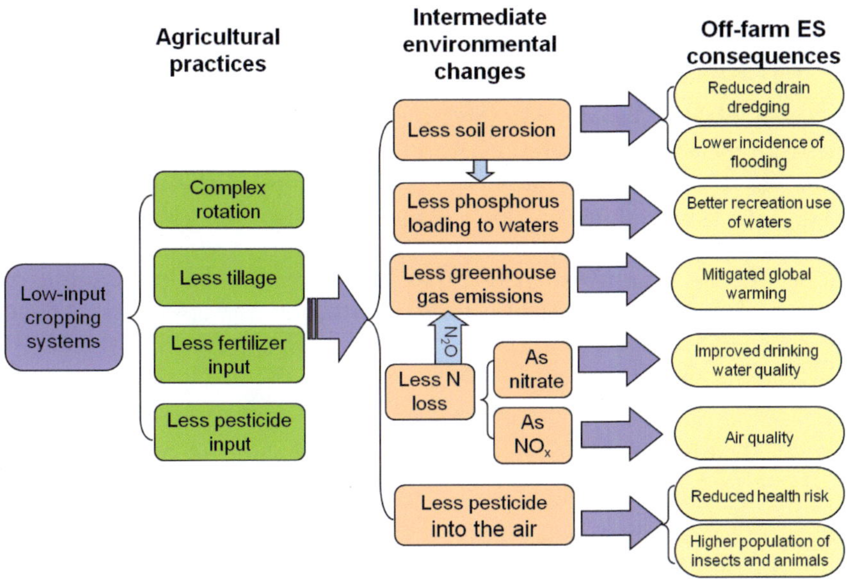

FIGURE 4-1 A schematic showing how farming practice changes link to environmental changes leading to outcomes experienced by consumers.
NOTE: ES = ecosystem services, N_2O = nitrous oxide, NO_x = nitrogen oxides.
SOURCE: Chen, 2010.

marketed foods. Finally, because ecosystem services vary in different settings and with different practices, they have to be measured separately in all of those different situations. That can be costly. While simulation models can help with some measurements, their validity has been tested only for a select range of applications.

IMPACT OF THE FOOD SYSTEM ON HEALTH INEQUALITIES[3]

"I think throughout this workshop, we should be saying not just 'cost' generically, but 'cost to whom,'" said Steve Wing. In Wing's opinion, many external costs to the food system are strongly related to health inequalities, which he defined as differences in morbidity, mortality, or health and well-being among people within a population or between populations. Wing described the various types of health inequalities that have been associated mostly with animal production, but also animal consumption; he argued that many external health costs associated with animal production are related to independence of the communities where CAFOs are located.

[3] This section summarizes the presentation of Steve Wing.

Health inequalities are often discussed within the context of racial groups that have experienced systematic institutional discrimination. According to Wing, health inequalities are persistent around the world and have been observed and quantified for centuries. In the United States, some health inequalities existed before the current industrial agricultural system existed and are therefore obviously not dependent on that system. However, in Wing's opinion, those inequalities are maintained in part by how the current food system operates. Many health inequalities related to food production stem from the fact that CAFOs tend to be located in low-income rural areas with disproportionate numbers of people of color who have experienced discrimination and already have poor health conditions for other reasons, including poverty.

Health Inequalities Associated with Animal Production

Wing identified three categories of health inequalities related to food production: (1) occupational or environmental health inequalities (i.e., the health and well-being of individuals); (2) built environment inequalities (i.e., aspects of the community that affect its functioning and the ability of a community to promote individual health and well-being); and (3) socio-political impact inequalities. Although he viewed the issue as mostly a rural (versus urban) issue now, he said it will increasingly become a global issue in the future. The CAFO production system is expanding globally and in places where communities may have even less capacity to push back against the negative impacts.

Occupational Health Inequalities

Occupational health inequalities arise from unsafe working conditions, low wages, and lack of supportive medical services (Lipscomb et al., 2005, 2006). Wing explained that many workers do not have access to medical services that are independent of the employer and often experience an unwillingness from employer-provided medical service personnel to assign a cause of injury or other medical condition to the working conditions because of implications for insurance and liability. Occupational health inequalities in the food system come from acute injuries, repetitive motion injuries, dermatological and respiratory conditions, psychosocial stress, infectious diseases, and other outcomes of exposures that occur in agriculture and food processing (Donham et al., 2000; Lipscomb et al., 2007a,b,c, 2008).

Contaminants that affect workers are also found in nearby communities. According to Wing, production-related pollutants (e.g., particulate matter, hydrogen sulfide) are present in neighborhoods located near

CAFOs and they are correlated with respiratory symptoms and lung function (Schinasi et al., 2011; Wing and Wolf, 2000), as well as quality of life, well-being, and mental health (Horton et al., 2009; Tajik et al., 2008; Wing and Wolf, 2000; Wing et al., 2008). Infectious diseases are another environmental health issue, due to water- and airborne pathogens. Furthermore, CAFO workers may serve as conduits for microbes—and antimicrobial resistance—to travel between the animal and human populations (Graham et al., 2008; Silbergeld et al., 2008). Wing said "good evidence" shows that workers act as conduits for the influenza virus to travel between animal and human populations (Gray and Baker, 2007; Gray et al., 2007; van Cleef et al., 2010). While human influenza pandemics emerging from animal populations are rare events, they have the potential to create extreme disruption and cost, not just to the animal industry but to the economy as a whole.

Built Environment Inequalities

Built environment inequalities relate mostly to housing value and well-being. For example, housing values may be impacted by the presence of offensive odors and the inability of residents to use their property (e.g., to garden, to recreate outside, to have family members visit). Not only do these effects impact the well-being of people who already live in the area, but they also make the built environment a less attractive place for others to live and discourage health-promoting and other community-building activity.

Sociopolitical Inequalities

Wing explained his view that sociopolitical inequalities are a consequence of animal production profits not being shared by local communities, which affects the ability of communities to promote their own health and well-being. Wing argued that many of the external costs of animal production are related to independence among the rural communities where CAFOs tend to be located. That is, a rural community does not always have self-determination to make its own decisions about land use, housing, and other community activities.

Health Inequalities Associated with Animal Consumption

Health inequalities are not just an animal production issue. They are also an animal consumption issue. Wing noted that many people who live in low-income areas have limited food choices, with some animal-based products being marketed to low-income people that are not being consumed by high-income people (see also the next section on accessibility to food). So while there are increasingly healthy, high-end foods on the market that are being produced in ways, for example, that reduce the potential growth of antibiotic-resistant bacteria, those foods are available only to people who can pay high prices for them. The remainder of the population does not have that choice. "I think we should be talking about the bifurcation of the food system," Wing said.

ACCESSIBILITY TO FOOD[4]

Accessibility to food is a social issue, claimed Ricardo Salvador. He described food insecurity on a global, national, and local (Washington, DC) level; explored the relationships among food insecurity, poverty, and health outcomes; and made the case that using economic analysis as a framework for addressing the full cost of food does not deal with the fundamental underlying issue of food accessibility.[5]

Food Insecurity

Consider food insecurity on a very macro scale, that is, a global scale, where 12.5 percent (870 million people) of the world population goes hungry. Then consider that the proportion of U.S. residents who are food insecure is greater than the global average, with 15.7 percent (48.8 million) in the United States going hungry. Then consider that, within Washington, DC, itself, the food insecurity rate is 13 percent (84,000 persons). Salvador noted there were six full-service, sit-down restaurants on the same block where the workshop was being held and another nine on contiguous blocks. "For those of us sitting here," he said, when we think about food, our thoughts tend to focus on time (e.g., "How much time do we have to eat?") and choice (e.g., "Do we want Thai or Mexican or . . . ?"), with the power of our income making food resources flow to us wherever we happen to be. "But that is not the reality for everyone on this planet," he said. People who are food insecure do not have that same power.

For example, a Washington, DC, map of all 34 full-service grocery stores (24 national chains dominated by Safeway, but also Giant, Whole Foods, and Trader Joe's, plus another 10 regional stores), shows that more than half are in the wealthiest part of the district, the Northwest section, and few are in what is primarily an African American area with a very high poverty rate (i.e., Wards 7 and 8). Salvador stated that the lack of full-service grocery stores in Wards 7 and 8 is not a function of the owners of

---

[4] This section summarizes the presentation of Ricardo J. Salvador.

[5] Salvador noted that although access to food can be categorized as either physical or economic, physical access is a subset of economic access.

those stores failing to notice that there is a population living on the east side of the Anacostia River. Rather, it is a function of the owners recognizing the lack of purchasing power in that population. People living in Wards 7 and 8 do not earn enough income to make it worthwhile for the owners to provide them with the same food system that serves "those of us sitting here."

During the past several decades, Americans' share of disposable personal income spent on food has decreased, from about 24 percent in 1930 to about 9.47 percent in 2010, based on U.S. Department of Agriculture data. The implication of this trend, Salvador observed, is that vast increases in productivity and efficiency have created a "very cheap" food supply. He pointed out, however, that the 9.47 percent figure is a ratio, with a numerator and a denominator. At least part of the decreasing trend in share of disposable income spent on food is a function of an increasing denominator and the fact that the average American income has increased tremendously over the past several decades. The absolute cost of food remains high. For those living in poverty, a much higher percentage of income is therefore spent on food. In a comparison across countries with variable average total household expenditures (e.g., $32,051 in the United States and $21,788 in the United Kingdom, compared to $620 in India and $541 in Kenya), the percentage of average total household expenditures spent on food increases as average household expenditures decrease (e.g., 6 percent in the United States and 9 percent in the United Kingdom, compared to 35 percent in India and 45 percent in Kenya).

The fact that food is not "very cheap" for people living in poverty has implications for food insecurity, Salvador implied. In the United States, a map of the percentage of people living in poverty areas overlays fairly well with a map of the percentage of households that are food insecure. For example, Mississippi is among those states with the highest percentage of its population living in poverty (i.e., 30 percent or more). It also has among the highest average rates of food insecurity.

In addition to food security implications, the fact that food is not very cheap for people living in poverty also has implications for health. Not only do U.S. poverty and food insecurity maps overlay, but both maps also overlay a map of adult obesity rates in the United States. This is because limited access to food limits the options available and the choices one can make. "The choices . . . are not going to be optimal," Salvador said. This is especially true for children who are too young to make any conscious choices at all. Salvador showed a photograph of two obese children eating a meal at McDonald's and observed, "They are simply a reflection of the food system that has been built up around them."

Obesity rates are increasing among wealthier Americans as well, according to Salvador. As with lower income Americans, wealthier Americans also reflect the food environment around them and the choices they can make. In that sense, Salvador suggested that merchants compete for their patronage by blending salt, sugar, and fat into textures and flavorings that make people want to eat at their restaurants. Extrapolated over one's lifetime, the negative health effects of eating those combinations of salt, sugar, and fat accumulate. Self-reported data from both Americans and Canadians show that obesity rates increase as age increases, until about the mid-60s, and that obesity rates for all age groups has increased during the past several decades. "These are consequences of actual access to food," Salvador said.

Implications of Food Insecurity for a Study on the Cost of Food

Salvador cautioned that the dominant global industrial food system that serves the wealthy excludes the reality of people who do not have access to that system. Discussions of the "hidden" costs of this food system, including the health and environmental impacts, excludes the reality of folks who do not participate in the system or who experience it in what Salvador described as "totally different ways." He said, "I think the economic analysis that we have been discussing predominantly over the last day necessarily has to inform what it is that we do when we talk about the actual price of food. But I also want to make the case that it is a constrained window into the broader issue that we need to be aware of in order to really do justice to the topic."

A Novel Approach to Addressing Lack of Access to Food

Salvador mentioned several programs and incentives that have attempted to address lack of access to food caused by poverty. Some of these programs and incentives have been publicly funded, others implemented through public-private partnerships. They include the Supplemental Nutrition Assistance Program (SNAP); School Breakfast Program (SBP); National School Lunch Program (NSLP); Special Supplemental Nutrition Program for Women, Infants, and Children (WIC) program; Farmers' Market Promotion Program (FMPP); EBT (Electronic Benefits Transfer)/SNAP redemptions at farmer's markets; Healthy Corner Stores Network (HCSN); fresh food financing; and food hubs.

Rather than creating systems such as these that provide ways for the poor to access the food system when hunger has become a reality, Salvador suggested shifting the food system so it can be accessed to prevent hunger and metabolic diseases from occurring. As an example of how this might be done, he mentioned healthy local food system models that acknowledge not only the key economic functions of a food system (e.g., production, processing, distribution, consumption) but also the values and social parameters that define how the food system functions (e.g., fairness, economic balance [i.e., all citizens have easy access to a variety of foods], sustainability, transparency, health promotion).

ANIMAL WELFARE[6]

Animal agriculture has gone through many changes in the past 30-40 years, resulting in about a doubling of meat production per sow—from about 2,000 pounds of pork per breeding sow in 1983, to nearly 4,000 in 2007. As a consequence of increased productivity, pork prices dropped 44 percent between 1973 and 2007. Beef and chicken prices dropped by 36 and 61 percent, respectively, over the same time period. But with these benefits come some concerns about the living conditions of the animals in the new meat production systems. Jayson Lusk discussed recently implemented animal welfare regulations; the potential costs and benefits of such regulations; and how to measure and value animal welfare.

Animal Welfare Legislation

Many animal welfare regulations are state level, mostly state ballot initiatives and, in some cases, state legislation. For example, several states have banned gestation crates[7] or battery cages[8] in agricultural production, including three of the top egg-producing states (California, Michigan, Ohio). According to Lusk, the increasing number of states that have banned battery cages has created a demand for more uniform, national legislation, with United Egg Producers and the Humane Society of the United States agreeing to push for a new national standard. In Lusk's opinion, a similar national push has yet to be observed in the pork industry because only one of the top pork-producing states, Ohio, has been affected by state-level animal welfare legislation (Corbin et al., 2012).

The Costs of Animal Welfare Regulation

Lusk identified several costs of animal welfare regulation to both producers and consumers. For example, a nationwide ban on battery cages would cost producers an estimated $187 million per year and a nationwide ban on gestation crates an estimated $258 million per year. The costs include the capital costs of switching from one system to another. For consumers, a battery cage ban would cost an estimated $1.8 billion per year as

[6] This section summarizes the presentation of Jayson Lusk.
[7] A sow stall used in pig farming.
[8] A cage used for egg-laying hens.

a result of a projected 21.18 percent price increase. A gestation crate ban would cost consumers an estimated $738 million per year as a result of a projected 1.72 percent price increase. Lusk speculated that price increases for eggs or pork would likely have a greater impact on poorer consumers. Just as previous speakers highlighted various inequalities, Lusk stressed concern about the burden on the poor if the cost of food is increased (i.e., the regressive nature of food taxation that attempts to curb food externalities). In addition to price changes, he identified less choice as another consumer cost. Right now, consumers have several egg options (e.g., cage-free eggs, organic eggs, omega-3 eggs). Changing the production process would remove some of the cheaper options. Another potential consumer cost is the likelihood of food shortages immediately after a ban, such as what occurred in the European Union following their ban on battery cages. The specific cost impact of such a ban may depend on the alternatives that remain after the ban has been enacted (i.e., a ban on all cages would have a different cost than a ban on a specific type of cage only).

In addition to producer and consumer costs, animal welfare legislation has environmental costs. According to Lusk, research indicates a somewhat lower feed efficiency in cage-free systems because animals exert more energy when they walk around, dust-bathe, etc. Additionally, often more land is required to produce the same volume of meat or eggs, which would increase the cost associated with the use of that land. While bans typically result in fewer animals being raised, Lusk said it is likely that both feed and land would increase on net.

Finally, in Lusk's opinion, there could even be a cost to some animals if a cost could be attached to an animal that would otherwise be brought into existence if not brought into existence because of a ban.

Benefits of Animal Welfare Regulation

Lusk identified several potential benefits to consumers and animals. For consumers, animal welfare legislation could fix an "information problem," that is, it could provide consumers with more knowledge about production conditions and would probably impact their purchasing choices among meat and egg products. Another potential benefit is the gain for individuals who do not approve of meat production and are impacted by other people's consumption choices.

The largest benefit of animal welfare legislation, in Lusk's opinion, is to the animals themselves. Most models indicate that animals experience a higher level of animal welfare in cage-free systems and exhibit more natural behaviors such as dust-bathing, flying on perches, and rooting. While there seems to be increased mortality in cage-free systems, there is controversy over how to value differences in mortality rates among different types of systems.

Measuring Animal Welfare

There are many models for measuring animal welfare. Lusk mentioned two: the SOWEL model (SOw WELfare; Bracke et al., 2002a,b) and the FOWEL model (FOwl WELfare; De Mol et al., 2006). Both models consider all the inputs (e.g., space, stocking density, feed, water), weight those inputs according to their relative importance with respect to affecting animal behavior, and then calculate and rank animal welfare scores. For example, Bracke et al. (2002a,b) used the SOWEL model to score and rank different pork production housing systems and reported that the individual stall system, which is the primary system used in the United States, ranks relatively low compared to the family pen and other systems. In Lusk's opinion, that does not mean that the U.S. pork industry should switch to the family pen or one of those other systems. There would be costs to doing so. But according to these models, the animals would be more comfortable in those other systems.

Valuing Animal Welfare

Lusk asked, "Can we value animal welfare?" He thinks the answer is "yes." The question is, how? One way is to examine actual cost differences (Figure 4-2). For example, scanner data on average egg prices in the

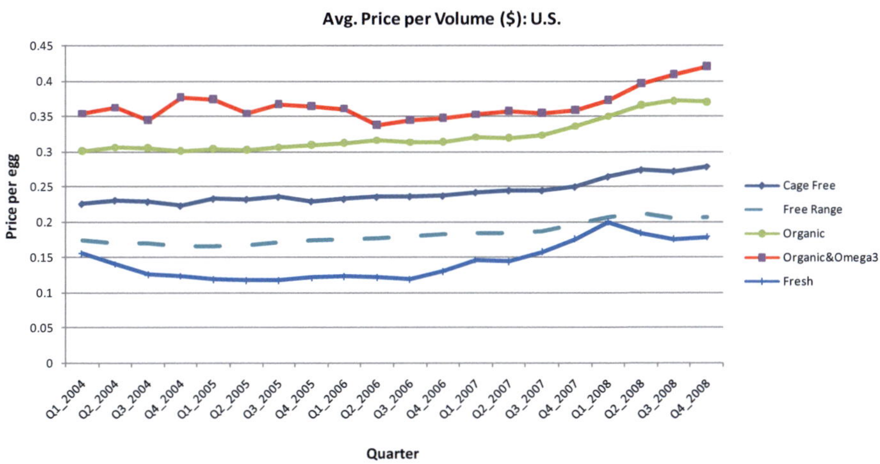

FIGURE 4-2 Cost differences across different types of eggs from 2004 to 2008.
SOURCE: Retail scanner data, IRI.

United States over a 4-year period show that, on average, organic eggs are about twice as expensive as "conventional" eggs. The higher price can be interpreted in two ways. Either it is costly to produce organic eggs, or people prefer organic eggs. Scanner data showing what people actually buy indicate that the market share for organic and other "specialty" eggs (e.g., cage free and organic, natural) increased over the same 4-year period, and the sum of all types of "specialty" eggs adds up to less than 5 percent of the market share. That very small market share suggests that most people are not willing to pay high prices for organic eggs (Figure 4-3).

A second way to value animal welfare is to conduct surveys and experiments. Most research suggests that consumers are willing to pay more for products produced in a certain way when they are informed of different production practices. However, it is unclear whether WTP for higher levels of animal welfare exceeds the cost of production.

In Lusk's opinion, much of the value that people derive from animal welfare is a personal value. With respect to public policy, the question is, Do these personal values impose externalities? It is not clear whether WTP for greater animal welfare reduces external costs to other humans. A more compelling case, Lusk said, is that WTP for greater animal welfare reduces the external costs for the animals themselves. After all, they are the ones who suffer. Lusk referred workshop participants to a recent paper for a

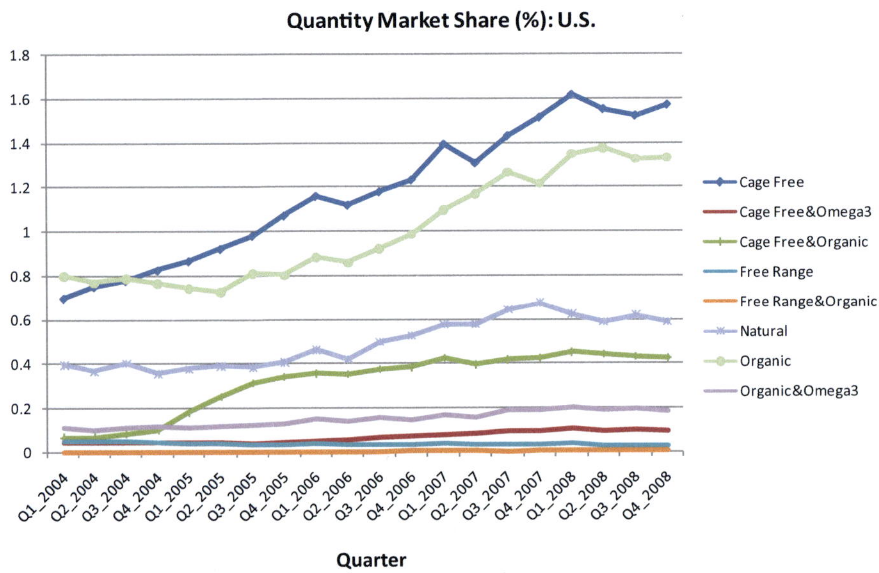

FIGURE 4-3 The percent market share of consumer egg purchases.
NOTE: The sum of all types is less than 5 percent.
SOURCE: Retail scanner data, IRI.

discussion of markets that allow for the buying and selling of improvements in animal welfare (Lusk, 2011).

The Importance of Trade-Offs

"*Energy is indeed a scarce and valuable resource, but it is only one of many, and there is a good deal more to life than British thermal units.*"
—Robert Dorfman (1977)

Lusk concluded by emphasizing the importance of keeping trade-offs in mind. The question, he said, is not the well-being of animals. The question is, what do we have to give up to get that benefit? For example, cages and other technological developments (e.g., feed additives) might decrease animal welfare, but they also improve some environmental outcomes. Beef cattle arguably experience the highest levels of animal welfare of any animal sources, yet beef production generates the largest negative environmental consequences (e.g., with respect to CO_2 emissions). Reducing CO_2 emissions would require a shift toward a system with greater animal welfare costs.

Another trade-off to consider, in Lusk's opinion, is that "meat tastes good." The costs associated with animal production reveal nothing about "how happy we are about that pound of meat we consumed." Of greater interest to Lusk than the amount of energy required to produce a pound of meat is the marginal utility or "the extra happiness" derived from that one unit of energy consumption. He said, "To me, those are the questions I think that are really fundamental. . . . It is not the fact that we use energy, it is what do we get out of the energy we use?"

REFERENCES

Bockstael, N., A. Freeman, R. Kopp, P. Portney, and V. K. Smith. 2000. On valuing nature. *Journal of Environmental Science and Technology* 34(8):1384-1389.

Bracke, M. B. M., B. M. Spruijt, J. H. M. Metz, and W. G. P. Schouten. 2002a. Decision support system for overall welfare assessment in pregnant sows. A: Model structure and weighting procedure. *Journal of Animal Science* 80:1819-1834.

Bracke, M. B. M., J. H. M. Metz, B. M. Spruijt, and W. G. P. Schouten. 2002b. Decision support system for overall welfare assessment in pregnant sows. B: Validation by expert opinion. *Journal of Animal Science* 80:1835-1845.

Chen, H. 2010. *Ecosystem services from low input cropping systems and the public's willingness to pay for them.* M.S. thesis. Agricultural, Food and Resource Economics. East Lansing, MI: Michigan State University.

Corbin, M., J. L. Lusk, and F. B. Norwood. 2012. *Nationwide stall ban likely?* Feedstuffs FoodLink. http://www.feedstuffsfoodlink.com/ME2/dirmod.asp?sid=&nm=&type=news&mod=News&mid=9A02E3B96F2A415ABC72CB5F516B4C10&tier=3&nid=CE5DBF418DF84BFDAAA4C97DFCF71331 (accessed October 19, 2012).

De Mol, R. M., W. G. P. Schouten, E. Evers, and H. Drost. 2006. A computer model for welfare assessment of poultry production systems for laying hens. *Netherlands Journal of Agricultural Science* 54:157-168.

Donham, K., D. Cumro, S. Reynolds, and J. Merchant. 2000. Dose-response relationships between occupational aerosol exposures and cross-shift declines of lung function in poultry workers: Recommendations for exposure limits. *Journal of Occupational and Environmental Medicine* 42:260-269.

Donham, K. J., S. Wing, D. Osterberg, J. L. Flora, C. Hodne, K. M. Thu, and P. S. Thorne. 2007. Community health and socioeconomic issues surrounding concentrated animal feeding operations. *Environmental Health Perspectives* 115:317-320.

Freeman, A. M., III, 2003. *The measurement of environmental and resource values: Theory and methods*. 2nd ed. Washington, DC: Resources for the Future.

Graham, J. P., J. H. Leibler, L. B. Price, J. M. Otte, D. U. Pfeiffer, T. Tiensin, et al. 2008. The animal-human interface and infectious disease in industrial food animal production: Rethinking biosecurity and biocontainment. *Public Health Reports* 123:282-299.

Gray, G. C., and W. S. Baker. 2007. The importance of including swine and poultry workers in influenza vaccination programs. *Clinical Pharmacology and Therapeutics* 82:638-641.

Gray, G. C., T. McCarthy, A. W. Capuano, S. F. Setterquist, C. W. Olsen, and M. C. Alavanja. 2007. Swine workers and swine influenza virus infections. *Emerging Infectious Diseases* 13:1871-1878.

Heal, G. 2000. *Nature and the marketplace: Capturing the value of the ecosystem*. Washington, DC: Island Press.

Horton, R. A., S. Wing, S. W. Marshall, and K. A. Brownley. 2009. Malodor as a trigger of stress and negative mood in neighbors of industrial hog operations. *American Journal of Public Health* 99(Suppl 3):S610-S615.

Lipscomb, H. J., R. Argue, M. A. McDonald, J. M. Dement, C. A. Epling, T. James, et al. 2005. Exploration of work and health disparities among black women employed in poultry processing in the rural south. *Environmental Health Perspectives* 113:1833-1840.

Lipscomb, H. J., D. Loomis, M. A. McDonald, R. A. Argue, and S. Wing. 2006. A conceptual model of work and health disparities in the United States. *International Journal of Health Services* 36:25-50.

Lipscomb, H. J., J. M. Dement, C. A. Epling, B. N. Gaynes, M. A. McDonald, and A. L. Schoenfisch. 2007a. Depressive symptoms among working women in rural North Carolina: A comparison of women in poultry processing and other low-wage jobs. *International Journal of Law and Psychiatry* 30:284-298.

Lipscomb, H. J., J. M. Dement, C. A. Epling, M. A. McDonald, and A. L. Schoenfisch. 2007b. Are we failing vulnerable workers? The case of black women in poultry processing in rural North Carolina. *New Solutions* 17:17-40.

Lipscomb, H. J., C. A. Epling, L. A. Pompeii, and J. M. Dement. 2007c. Musculoskeletal symptoms among poultry processing workers and a community comparison group: Black women in low-wage jobs in the rural south. *American Journal of Industrial Medicine* 50:327-338.

Lipscomb, H., K. Kucera, C. Epling, and J. Dement. 2008. Upper extremity musculoskeletal symptoms and disorders among a cohort of women employed in poultry processing. *American Journal of Industrial Medicine* 51:24-36.

Lusk, J. L. 2011. The market for animal welfare. *Agricultural and Human Values* 28:561-575.

MA (Millennium Ecosystem Assessment). 2003. *Ecosystems and well-being: A framework for assessment*. Washington, DC: Island Press.

Mirabelli, M. C., S. Wing, S. W. Marshall, and T. C. Wilcosky. 2006. Asthma symptoms among adolescents who attend public schools that are located near confined swine feeding operations. *Pediatrics* 118:e66-e75.

Pearce, D. 1998. Auditing the earth. *Environment* 40(2):23-28.

Schinasi, L., R. A. Horton, V. T. Guidry, S. Wing, S. W. Marshall, and K. B. Morland. 2011. Air pollution, lung function, and physical symptoms in communities near concentrated swine feeding operations. *Epidemiology* 22:208-215.

Silbergeld, E. K., J. Graham, and L. B. Price. 2008. Industrial food animal production, antimicrobial resistance, and human health. *Annual Review of Public Health* 29:151-169.

Swinton, S. M., F. Lupi, G. P. Robertson, and S. K. Hamilton. 2007. Ecosystem services and agriculture: Cultivating agricultural ecosystems for diverse benefits. *Ecological Economics* 64(2):245-252.

Tajik, M., N. Muhammad, A. Lowman, K. Thu, S. Wing, and G. Grant. 2008. Impact of odor from industrial hog operations on daily living activities. *New Solutions* 18:193-205.

van Cleef, B. A., E. J. Verkade, M. W. Wulf, A. G. Buiting, A. Voss, and X. W. Huijsdens, et al. 2010. Prevalence of livestock-associated MRSA in communities with high pig-densities in the Netherlands. *PLOS One* 5:e9385.

Wing, S., and S. Wolf. 2000. Intensive livestock operations, health, and quality of life among eastern North Carolina residents. *Environmental Health Perspectives* 108:233-238.

Wing, S., R. A. Horton, S. W. Marshall, K. Thu, M. Tajik, and L. Schinasi, et al. 2008. Air pollution and odor in communities near industrial swine operations. *Environmental Health Perspectives* 116:1362-1368.

5

Attaching Value to Costs and Benefits

Added to the challenges of identifying and then quantifying environmental, health, and other impacts of the food system is the challenge of attaching monetary value to those impacts. Two speakers were invited to share their thoughts on the methods and challenges of valuation. This chapter summarizes their presentations. (See also Chapter 3 for a summary of James Hammitt's explanation of how economists attach value using either willingness to pay, or WTP, or another measure known as the quality-adjusted life year, and how both valuation methods are challenged by social aggregation problems.)

Based on his experience as a member of the National Research Council (NRC) committee that defined and evaluated the external costs and benefits associated with the production, distribution, and consumption of energy (NRC, 2010), James Hammitt discussed challenges to quantifying and monetizing external costs and benefits and issues to consider when conducting a similar study on the cost of food. Anna Alberini, associate professor of economics in the Department of Agricultural and Resource Economics at the University of Maryland, College Park, discussed how economists attach value to external costs and benefits using estimates of WTP and "value of a statistical life" (VSL), and identified several factors to consider before transferring estimates of monetary value from one context to another (e.g., using VSL estimates obtained in manufacturing or other nonfood contexts in a study on the cost of food). She cautioned that monetizing the costs and benefits of food will probably require multiple valuation exercises, with different effects considered separately.

LESSONS FROM *THE HIDDEN COSTS OF ENERGY: UNPRICED CONSEQUENCES OF ENERGY PRODUCTION AND USE*[1]

In 2010, the NRC released a report on the externalities of energy production and use (NRC, 2010). The study was requested by Congress in the Energy Policy Act of 2005, with funds appropriated to the U.S. Treasury in the Consolidated Appropriations Act of 2008. According to Hammitt, the study was conducted under a very tight time schedule, with the first of six committee meetings held on September 11, 2008, and a 473-page report released on October 19, 2009. Hammitt discussed major challenges to completing the statement of task for that report; described key results; and highlighted questions to consider for a study on the cost of food.

Statement of Task

Key components of the statement of the task were to (italics added by Hammitt for emphasis): "*define and evaluate key external costs and benefits* . . . associated with the *production, distribution and consumption of energy* from various selected sources *that are not or may not be fully incorporated into the market price* of such energy" and to "carry out its task from a *U.S. perspective*," but also "*consider broader geographic implications of externalities when warranted and feasible.*" Among other activities, Congress requested that the committee "*identify key externalities* . . . in the categories of *human health*, *environment*, *security* (including quality, abundance, and reliability of energy sources), and *infrastructure* (such as transportation and waste disposal systems not sufficiently taken into account by producers or consumers)"; "consider *externalities associated with . . . energy imported from foreign sources*"; "develop an approach for estimating externalities related to greenhouse gas emissions and *climate change*"; and "*present qualitative* and, to the extent practicable, *quantitative estimates of externalities and associated uncertainties.*" (See Box 5-1 for the NRC, 2010, definition of externality.)

Major Challenges

Hammitt discussed three major challenges faced by the NRC (2010) committee: (1) identifying internalized externalities; (2) quantifying and monetizing all endpoints; and (3) exploring the disproportionate amount of effort focused on already well-understood externalities.

[1] This section summarizes the second presentation of James K. Hammitt.

BOX 5-1
The NRC (2010) Definition of Externality

The NRC (2010) report defined externality as follows: "an externality, which can be positive or negative, is an activity of one agent (for example, an individual or an organization, such as a company) that affects the well-being of another agent and occurs outside the market mechanism." Assuming that people respond to prices and nothing else, a logical extension of that definition is, as stated in the report, "In the absence of government interaction, externalities associated with energy production and use are generally not taken into account in decision making."

SOURCE: NRC, 2010, p. 29.

Identifying Internalized Externalities

One key component of the statement of task was to evaluate key externalities "that are not or may not be fully incorporated into the market price . . . or into the federal tax or fee." Without knowing whether an externality has been internalized or not, it is impossible to know whether a policy change could improve the situation. According to Hammitt, this component of the task was especially challenging because of the difficulty in determining, in some cases, whether an externality is internalized or not. It is not always clear.

For example, the following scenario was used in the report to illustrate the concept of externality: "A coal-fired electricity-generating plant, which is in compliance with current environmental regulations, releases various pollutants. . . . The damage from this pollution is . . . a 'social cost.' If these social costs were not adequately taken into account in selecting the plant's site or the air pollution control technology that it uses, the true costs . . . have not been reflected in these decisions." Hammitt observed that while damage from this pollution is clearly a social cost, whether that cost has been internalized is not clear. If the social cost was not considered during selection of the plant location or when deciding which air pollution control technology to use, then it has not been internalized. But how does one know if that cost was adequately taken into account at the time those decisions were made? That it is a very difficult phenomenon for a committee to judge. Compounding the challenge is the likelihood that science has evolved since that time, so knowledge about harm from pollution is different than it was when those decisions were made. So in that case, the committee was able to estimate both total damages (compared with zero emissions) and marginal damages (the damage that arises from the last unit of emission or other type of burden). But without knowing whether those costs had been accounted for during the decision making about location, technology, etc., the committee was unable to judge whether the externalities had been internalized or not.

Quantifying and Monetizing All Endpoints

The scope of the study spanned across a range of major or rapidly growing energy sources and carriers, with major endpoints being human health, climate change, and infrastructure and security. Although environmental and ecological endpoints were part of the committee charge, the lack of data and good analytical frameworks for evaluating those endpoints made it impossible for the committee to quantify or monetize them.

Exploring All Externalities

Hammitt's personal view on the scope of the study was that too much effort was focused on well-understood damages—that is, quantifying basic human health effects associated with fossil-fuel combustion—and too little effort focused on describing other externalities. He said, "We had the opportunity here to try and do something more innovative—less quantitative, but potentially pushing the field further along." For example, in Hammitt's opinion, there could have been more effort directed toward describing security and infrastructure, or unconventional power (wind, solar, etc.). There was also some inconsistency across sectors, with health damages from fossil fuels being quantified, but health damages from infrastructure and security not being quantified. Arguably, many of the external costs associated with infrastructure and security are either too difficult to quantify or already fully internalized. For example, many attack scenarios (i.e., attacks on facilities) are internalized through corporate liability and other measures. With respect to dependence on foreign oil, which was explicitly in the charge, there might be some costs associated with military activity in oil-producing regions of the world. But those constraints are difficult to estimate, and the marginal effects of U.S. oil consumption on those activities may be negligible.

Use of Graphics

Graphics used in the energy report included flow charts showing which elements of the system were examined; pie charts and bar graphs showing consumption by source and use by sectors; and tables showing which components of the system were examined using quantitative versus qualitative methodologies (see Figure 5-1a-d). Hammitt referred to Heller and

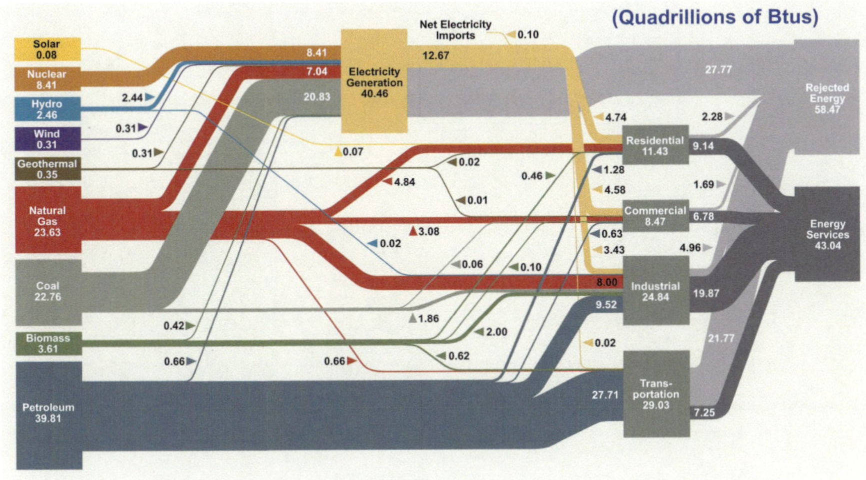

(a) Energy flows in the U.S. economy, 2007

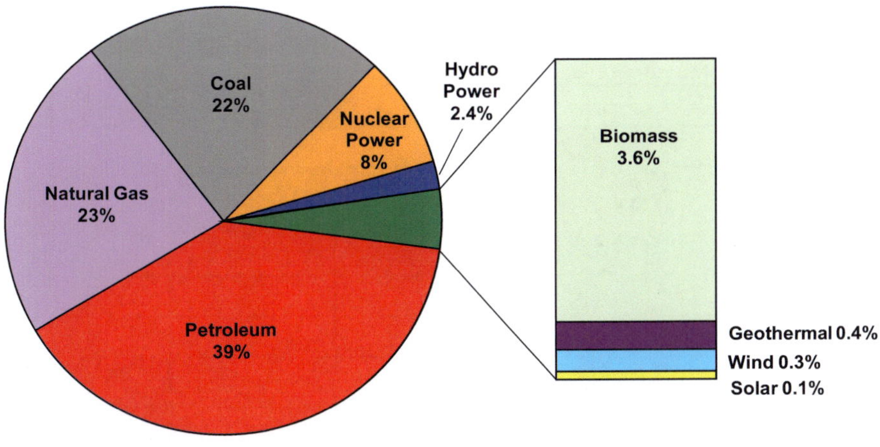

(b) 2007 U.S. energy consumption by energy source. Total consumption = 101.5 quadrillion Btu

FIGURE 5-1 Examples of graphics used in *The Hidden Costs of Energy*: (a) the flow of major energy sources and uses; (b) energy consumption by energy source; (c) primary energy use by sector; (d) analytical methods used to examine various source/sector combinations.

NOTE: The focus of the energy report was on four major carriers or users of energy: electricity (Chapter 2), transportation (Chapter 3), industrial heat (Chapter 4), and

continued

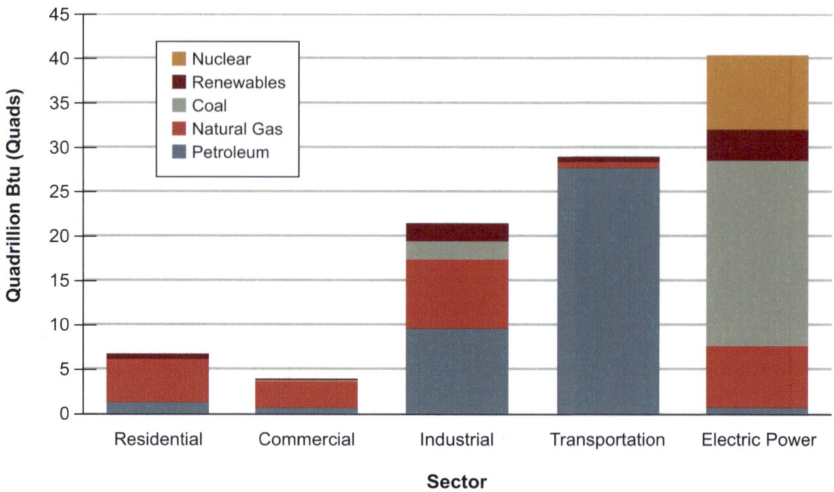

(c) U.S. primary energy use by sectors (2007)

Energy Source	Electricity (Chapter 2)	Transportation (Chapter 3)	Industry—heat (Chapter 4)	Buildings—heat (Chapter 4)
Oil		MA	QE	QE
Coal	MA			
Natural Gas/Liq	MA	MA	QE	MA
Uranium	QL			
Biomass	QL	MA		
Wind	QL			
Solar Power	QL			QE
Other Fuels		MA		
Electricity	-	MA	QE	QE

(d) Methods used to examine source/sector combinations

FIGURE 5-1 Continued
commercial/residential heat (Chapter 4). Even though climate change is associated with all of those carriers/users, it was treated separately (Chapter 5), as was infrastructure and security (Chapter 6). Btu = British thermal unit, MA = quantitative modeling analysis conducted by the committee, QE = qualitative evaluation, QL = quantitative information from the literature.
SOURCE: NRC, 2010.

Keoleian's (2003) schematic of material flow of the U.S. food system as a similar kind of graphic that might be useful for communicating results of a study on the cost of food (see Figure 5-2).

Select Results of the NRC (2010) Analysis of the External Costs of Energy

Hammitt highlighted two key sets of findings from the NRC (2010) study. First, the committee's analysis of electricity, which was based partly on a detailed modeling of air pollution mortality, revealed that the dominant outcome is particulate matter mortality associated with coal and natural gas plants, but with a great deal of heterogeneity among plants. Average damage among the 406 coal-fired electric plants across the United States is $160 million per plant, but with plants in the top decile causing an average $666 million per plant (in 2007). It was likewise with natural gas plants. Average damage among the 498 gas plants in the United States is $1.5 million per plant, but with plants in the top decile causing $9.73 million per plant on average (in 2007). The heterogeneity is not just per plant, but also geographic, with most of the higher decile coal plants (i.e., those causing the

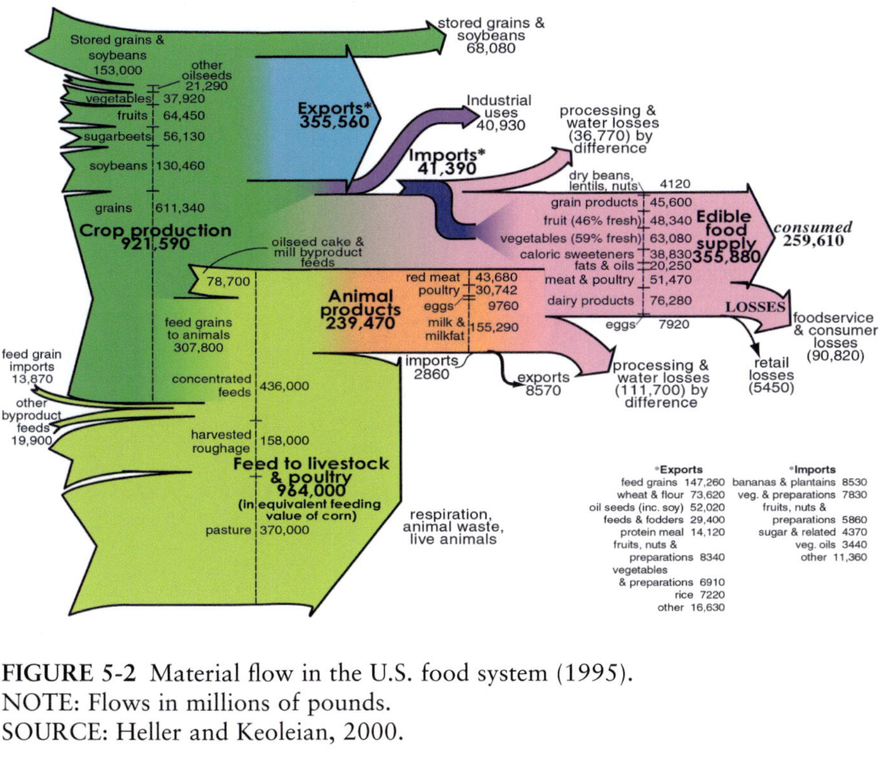

FIGURE 5-2 Material flow in the U.S. food system (1995).
NOTE: Flows in millions of pounds.
SOURCE: Heller and Keoleian, 2000.

most damage) in the Ohio Valley and most of the higher decile natural gas plants on the west coast, in Texas, and along the eastern seaboard.

Second, the committee's analysis of on-road transportation, which involved using life cycle analysis to examine a wide range of fuels and engine technologies, revealed that total damages were remarkably similar across different fuel/engine technologies. For example, health damages caused by light-duty automobiles were nearly 1.5 cents per vehicle mile traveled for almost all fuel/engine technologies, with liquid hydrogen being the only outlier. It was likewise with greenhouse gas (GHG) emissions associated with light-duty automobiles. Again, there was not much difference among various types of fuel/engine technologies, except with biofuels (dry corn, wet corn, herbaceous, and corn stover), where some withdrawal of carbon from the atmosphere offsets emission. "This is pretty interesting," Hammitt said, "to have these 20 different technologies and such small differences among them."

A summary of the monetized health and climate effects for electricity and on-road transportation are shown in Table 5-1, along with the health and climate effects of heat. Hammitt highlighted the "big" effects: damages from coal-fired electricity, at $62 billion per year; damages from light-duty transportation, at $36 billion per year; and damages from heavy-duty transportation (i.e., trucks), at $20 billion per year.

TABLE 5-1 Monetized Health and Climate Effects Associated with Electricity, On-Road Transportation, and Health, Based on the 2010 (NRC) Report on the External Costs of Energy

Source/Use of Energy	Total Damage per Year	Health Damage per Unit	Climate Damage per Unit
Coal-fired electricity	$62 billion	3.2[<0.5-12] cents/kwh	3[1-10]cents/kwh
Natural gas-generated electricity	$0.74 billion	0.16[<0.05-1] cents/kwh	1.5[0.5-5]cents/kwh
Light-duty on-road transportation	$36 billion	1.2-1.7 cents/VMT	0.5[0.05-5]cents/VMT
Heavy-duty on-road transportation (trucks)	$20 billion		
Natural gas–generated heat	$1.4 billion	11 cents/MCF	0.7[0.07-7]cents/MCF

NOTE: kwh = kilowatt hour; MCF = 1,000 cubic feet; VMT = vehicle mile traveled. The numbers in brackets are uncertainty ranges.
SOURCE: NRC, 2010.

Questions to Consider When Planning a Study on the Cost of Food

The NRC (2010) study on the external costs of energy raises several questions to consider when planning a study on the external costs of food:

- Is the goal to estimate total or marginal externalities? Hammitt suggested analyzing both total and marginal effects. Both are useful for different reasons.
- Will an effort be made to deal with the extent to which externalities have been internalized? Hammitt cautioned that determining the extent to which externalities have been internalized can be especially challenging and suggested avoiding the challenge altogether if possible.
- Should pecuniary externalities be included in the analysis? Pecuniary externalities are effects that are transfers of income, or costs, between different parties. They do not impact total welfare, but they do impact distribution.
- Should the analysis consider externalities associated with foods that are imported from foreign sources? A related question is, should the analysis incorporate externalities that occur elsewhere but that harm the United States, such as GHG emissions and habitat/biodiversity loss?
- To what extent should the analysis consider heterogeneity of effects by location and other factors?

VALUING AGRICULTURAL EXTERNALITIES AND PUBLIC HEALTH IMPACTS[2]

The notion of valuing the many different types of health effects associated with food production and consumption is "complex and complicated," Anna Alberini stated. The notion is this: If something is important to you, then you should be willing to pay for it—either to obtain it or, if it is something that is causing a negative effect, to get rid of it. Alberini discussed how economists attach value using WTP assessments; how economists quantify WTP for a marginal change in mortality risk (i.e., VSL); and factors to consider before using VSL and other WTP estimates typically measured in nonfood contexts in a study on the costs of food.

Willingness to Pay

WTP is the maximum amount of money that an individual would voluntarily exchange to obtain an improvement or avoid an undesirable

[2] This section summarizes the presentation of Anna Alberini.

outcome. When evaluating the benefits of a policy, economists consider the sum of all beneficiaries' WTP for that benefit. Payment can be made in many ways, including via higher taxes, higher food prices, or by incurring costs or changing behaviors in order to protect oneself from the risk(s).

The simplest food-related example is the willingness to pay to reduce the risk of minor food poisoning or another acute illness. Economic theory indicates that the willingness to pay to avoid that risk is a function of several factors, including the medical expenditure to alleviate the symptoms that were actually experienced, any income that was lost to that illness (e.g., if a person did not have any sick days but was too sick to work), the cost of averting illness, and the value of avoiding the discomfort of being sick. According to Alberini, most public health and food safety policy is based on a cost-of-illness approach that considers only the first two factors, and it is unclear to what extent considering the additional two components would impact total WTP. She suspects that consideration of those two additional components could increase total WTP as much as twofold. "So we are talking about relatively large numbers," she said, "even for relatively simple episodes."

At the other extreme are mortality risks. When estimating WTP to reduce a mortality risk, economists typically use either the VSL or the value of a statistical life year (VOLY). VSL is the willingness to pay for a *small* change in the risk of dying. Alberini emphasized that the change is indeed small. VSL estimates are widely used by many U.S. agencies. For example, in 1999, the U.S. Environmental Protection Agency (EPA) calculated VSL as $6.2 million. Adjusted for the cost of inflation, that amounts to about $8 million today. The U.S. Department of Transportation uses a smaller figure within the $3.7-$5 million range.

VSL is estimated using any of a number of empirical methods (see "Estimating VSL" section below). VOLY is a derived estimate and is usually inferred from VSL. Alberini said she was aware of only two studies where researchers tried to estimate VOLY from empirical data (Chilton et al., 2002; NEEDS, 2006). In Alberini's opinion, both of those studies were so problematic that the estimates are unreliable. The challenge with empirically estimating VOLY stems from the difficulty in asking people whether they would be willing to pay for a gain in life expectancy. It is unclear whether the question can be conveyed effectively and whether people really understand what they are valuing when they answer the question.

Estimating VSL

Estimates of VSL are based on empirical data, using any of a number of different approaches. Alberini listed four major types of approaches: (1) compensating wage studies (e.g., Viscusi, 1993; Viscusi and Aldy, 2003); (2)

consumer behavior studies where one measures expenditures on safety devices (e.g., a bicycle helmet that will reduce the risk of dying) (Jenkins et al., 2001) or observes the tradeoffs people make to reduce risks (Blomquist, 2004); (3) housing price hedonics (e.g., Gayer et al., 2000, 2002); and (4) stated preference methods (e.g., Alberini et al., 2007; Krupnick et al., 2002; Tsuge et al., 2005).

Compensating Wage Studies

Compensating wage studies involve collecting data on wage rates and everything that might be a determination of that compensation (e.g., age, experience, type of profession) plus the risk of a fatal accident and the risk of a nonfatal injury on the job. The assumption is that people are compensated for taking riskier jobs, that workers know the risks, and that the researchers are measuring those risks correctly. While many U.S. agencies use this approach to estimate VSL, Alberini opined that it is "hard to believe" that all of the assumptions are met. She referred to Steve Wing's presentation on health inequalities and the unlikelihood that workers in certain types of animal processing facilities are actually compensated more for working in those high-risk conditions. In fact, Alberini argued that it is probably the opposite, that is, those workers are probably paid less for taking riskier jobs, partly because they are not aware of the risks. There are a number of examples where VSL calculated using this approach is actually negative or insignificant. Also, compensating wage studies are typically conducted for the manufacturing industry, not food, and many researchers would be uncomfortable using the approach to evaluate food system risks.

Housing Price Hedonics

Housing price hedonics is similar to the approach that real estate agents use when estimating the value of a home. It involves regressing the price of homes on structural characteristics (e.g., square footage, number of floors), neighborhood characteristics, and the risk of experiencing a negative outcome in the area (e.g., as a result of exposure to pollution). VSL estimates derived from housing price hedonics are typically comparable to those estimated using compensating wage values. But again, the assumption is that the risks are known and that the researcher is measuring them correctly. For example, many neighborhood characteristics impact risk but have no data.

As with compensative wage studies, housing price hedonics is an approach not typically used with food. The hedonic pricing approach, however, could be used for food. For example, one could regress the price of a certain type of egg on characteristics of the egg and the risks associated with the production and consumption of that type of egg (e.g., risk of mortality from a foodborne infection, risk of cancer associated with something in the animal feed, climate change effects). However, again, one of the assumptions is that willingness to pay that price is based on people being aware of all the risks. Also, too many of the risks are correlated, making it difficult to disentangle the contribution of each. Alberini said, "It would be interesting if we could actually inform people about the different types of risks associated with the different types of eggs and let them choose the [type of egg] and indicate how much they are willing to pay for them."

Stated Preferences

The stated preferences method involves asking people how much they would be willing to pay for a reduced risk of dying by asking them to choose between two alternative interventions for reducing risk. For example, in one of her studies, Alberini and colleagues asked respondents which of two interventions they preferred: (1) a nationwide public intervention that reduces the risk of death from road traffic accidents (by 3 in 10,000 over 5 years) and costs each household 300 euros; or (2) a private intervention that reduces the risk of death from cancer (by 2 in 10,000 over 5 years) and costs each household 200 euros (Alberini and Scasny, 2011). One of the concerns with conducting a study like this is making sure that respondents understand the magnitude of the risks. There are visual tools that can be used to educate respondents about risks (e.g., Corso et al., 2001).

The Challenge of Transferring Value

A key question to consider when planning a study that involves valuing the environmental or public health costs of food is whether VSL estimates from other (nonfood) contexts can be used. "I don't have an answer for you," Alberini said. Researchers have expressed concerns about such practice, but there is little empirical work documenting resons for doing or not doing such "transfers."

To further elaborate, several factors affect the WTP for a mortality risk reduction and, therefore, whether estimates of VSL measured in one setting are appropriate for use in another setting. The same is true of WTP for less extreme health risks. First, are the beneficiaries comparable? Most work environments involve males ages 30 to 40. It is unclear whether VSL estimates based on that context can be extrapolated to the rest of the population. Second, is the nature of the risk comparable? For example, the risk of cancer is a risk that comes with much dread and with suffering attached to both the disease and the treatment. The mental anxiety associated with cancer could be such that a person is willing to pay much more to reduce their risk of dying from cancer compared to the risk of dying of something else, such as an accidental death. Third, how does WTP for a reduced risk compare between a risk reduction experienced now versus later? Fourth, are there competing risks (e.g., Eeckhoudt and Hammitt, 2001; Evans and Smith, 2006)? If a person thinks that a particular risk is a very small portion of the overall risk of dying, they may not be willing to pay as much to reduce that risk. Or, if a person thinks that a particular risk is not within their control (e.g., cardiovascular disease associated with exposure to air pollution, as opposed to cardiovascular disease associated with diet), his or her WTP may be different to reduce that same risk. Fifth, are there public programs in place to make it easier for private behavior to reduce a risk, which has been shown in theory to influence WTP (Shogren, 1990)? Finally, when thinking about risk, most people don't just think about the probability of the adverse event. They also think about a number of other attributes that can impact WTP for the same risk reduction (the immediacy of the effect, future generations, etc.) (Slovic, 1987).

Although the focus of her presentation was on health, Alberini said there are several approaches to valuing environmental effects. Economists favor what is known as the damage function approach, which involves quantifying the physical effects and then attaching a value to those effects. In addition to the methods discussed above, the monetary value of the effects can also be estimated using the travel cost method (a method that infers the value people place on visiting some site, generally for recreational purposes), the hedonic housing price method, and stated preference and other stated preference methods. Some of these methods are well suited to estimating the effects of food production practices on ecological systems, but do not lend themselves to valuing the human health effects of food production practices or safety levels.

In conclusion, Alberini encouraged valuation of the health and environmental effects of food production, but emphasized that a single valuation exercise is unlikely to be sufficient. Different effects will likely require different methods. "We are probably better off dividing up the chore into different tasks and facing them separately," she explained.

In the question-and-answer period following her presentation, Alberini remarked that WTP includes ability to pay. That is, people are willing to pay only what they can pay. She also mentioned that altruism is another understudied topic, that is, the willingness to pay for other people. She mentioned current focus group research on altruism being conducted by the EPA and other scientists.

REFERENCES

Alberini, A., and M. Šcasný. 2011. Context and the VSL: Evidence from a stated preference study in Italy and the Czech Republic. *Environmental and Resource Economics* 49(4):511-538.

Alberini, A., A. Longo, and M. Veronesi. 2007. Valuing environmental amenities using stated choice studies. In *Basic statistical models for stated choice studies, Vol. 8, The economics of non-market goods and resources*, edited by B. J. Kanninen. The Netherlands: Springer. Pp. 203-227.

Blomquist, G. 2004. Self-protection and averting behavior, values of statistical lives, and benefit cost analysis of environmental policy. *Review of Economics of the Household* 2(1):89-110.

Chilton, S., J. Covey, L. Hopkins, M. Jones-Lee, G. Loomes, N. Pidgeon, and A. Spencer. 2002. Public perceptions of risk and preference-based values of safety. *Journal of Risk and Uncertainty* 25(3):211-232.

Corso, P. S., J. K. Hammitt, and J. D. Graham. 2001. Valuing mortality-risk reduction: Using visual aids to improve the validity of contingent valuation. *Journal of Risk and Uncertainty* 23(2):165-184.

Eeckhoudt, L. R., and J. K. Hammitt. 2001. Background risks and the value of a statistical life. *Journal of Risk and Uncertainty* 23(3):261-279.

Evans, M. F., and V. K. Smith. 2006. Do we really understand that age-VSL relationship? *Resource and Energy Economics* 28(3):242-261.

Gayer, T., J. T. Hamilton, and W. K. Viscusi. 2000. Private values of risk tradeoffs at Superfund sites: Housing market evidence on learning about risk. *Review of Economics and Statistics* 82(3):439-451.

Gayer, T., J. T. Hamilton, and W. K. Viscusi. 2002. The market value of reducing cancer risk: Hedonic housing prices with changing information. *Southern Economic Journal* 69(2):266-289.

Heller, M., and G. Keolian. 2000. *Life cycle-based sustainability indicators for assessment of the U.S. food system*, Center for Sustainable System, Ann Arbor, MI. Report No. CSS00-04.

Heller, M., and G. Keoleian. 2003. Assessing the sustainability of the U.S. food system: A life cycle perspective. *Agricultural Systems* 76:1007-1041.

Jenkins, N. R., N. Owens, and E. Wiggins. 2001. Valuing reduced risks to children: The case of bicycle safety helmets. *Contemporary Economic Policy* 19(4):397-408

Krupnick, A., A. Alberini, M. Cropper, N. Simon, B. O'Brien, R. Goeree, and M. Heintzelman. 2002. Age, health and the willingness to pay for mortality risk reductions: A contingent valuation survey of Ontario residents. *Journal of Risk and Uncertainty* 24(2):161-186.

NEEDS (New Energy Externalities Developments for Sustainability). 2006. *Final report on the monetary valuation of mortality and morbidity risks from air pollution.* http://www.needs-project.org/RS1b/NEEDS_RS1b_D6.7.pdf (accessed October 19, 2012).

NRC (National Research Council). 2010. *Hidden costs of energy: Unpriced consequences of energy production and use.* Washington, DC: The National Academies Press.

Shogren, J. F. 1990. The impact of self-protection and self-insurance on individual response to risk. *Journal of Risk and Uncertainty* 3:191-204.

Slovic P. 1987. Perception of risk. *Science* 236:280-285.

Tsuge, T., A. Kishimoto, and K. Takeuchi. 2005. A choice experiment approach to the valuation of mortality. *Journal of Risk and Uncertainty* 31(1):73-95.

Viscusi, W. K. 1993. The value of risks to life and health. *Journal of Economic Literature* 31:1912-1946.

Viscusi, W. K., and J. E. Aldy. 2003. The value of a statistical life: A critical review of market estimates throughout the world. *Journal of Risk and Uncertainty* 27(1):5-76.

6

Exploring Costs and Benefits

This workshop emphasized small-group discussion as well as addition to speaker presentations to gain a greater understanding of the key issues regarding the measurement of costs and benefits of the food system through in-depth expert discussions on focused topics. The complete second half of the first day was spent with participants divided into working groups, with each working group focused on one of four categories of health or environmental effects: (1) energy and greenhouse gas (GHG) emissions; (2) soil, water, and other environmental consequences; (3) health consequences of antimicrobial use in agriculture; and (4) other public health consequences. Although the working groups focused distinctly on these categories, several participants highlighted the underlying complex linkages and interactions between them. For example, concentrated animal feeding operations may provide improved control of some pathogens of public health concern and create less waste per pound of product, but these large operations also produce considerable quantities of manure that may lead to run-off problems when applied to nearby cropland. To reiterate the linkages and provide a more complete picture, some effects may appear in more than one of the subsequent working group summaries.

Each working group was given a matrix worksheet (Table 6-1) that focused on the following six key items for consideration:

1. The source(s) of the effect
2. Whether the effect is an environmental, public health, or other type of effect
3. Methodologies and limitations to measuring the effect

TABLE 6-1 Explanation of the Matrix Provided to Working Groups to Report on Effects of Practices Spanning the Life Cycle of Foods

Source(s) of Effect	Environmental Effect	Measurement/ Limitations	Externality?	Public Health or Other Effect
What practice or action is the primary source of the effect? For example, are the effects from crop production, manure management, fertilizer use, from use of farm machinery, transportation, or dietary intake?	Describe the effects of practices that have an impact (positive or negative) on the environment.	What indicators or methods can be used to measure this effect? What methodological limitations may inhibit measuring this effect?	Is the effect an externality? Yes (Y), No (N), or Unclear (U)	Describe the effects of practices that have a public health, economic, or social impact (positive or negative). If this effect is an indirect or mediated effect of an environmental effect, list the related effects in the same row. If this effect is not indirectly linked to an environmental effect, start a new row.

4. Whether the effect is an externality
5. Trade-offs related to the source of the effect and methodologies and limitations to measuring trade-offs
6. The life cycle stage during which the effect occurs (e.g., production, processing, distribution)

The groups took different approaches to addressing these items, with some participants noting that the matrix did not allow for an exhaustive examination since components, like the magnitude of the effect, are missing. This chapter summarizes the discussions that took place during the small-group discussions.[1]

[1] The summaries of the working group discussions are intended to demonstrate the diversity of perspectives and divergent opinions and should not be construed as reflecting any group consensus.

Measurement/ Limitations	Externality?	Trade-offs Related to Alternative Strategies	Measurement/ Limitations	Life Cycle Stage
What indicators or methods can be used to measure this effect? What methodological limitations may inhibit measuring this effect?	Is the effect an externality? Yes (Y), No (N), or Unclear (U)	What are the trade-offs (economic, environmental, health, or other) associated with the practice(s) that are causing the effect?	What indicators or method can be used to measure the trade-offs? What methodological limitations may inhibit measuring the trade-offs?	At what stage of the life cycle is the source of this effect occurring? Choose one of the following: • resource origin • agricultural production • food processing, packaging & distribution • preparation & consumption

EFFECTS OF FOOD PRODUCTION, PROCESSING, AND CONSUMPTION ON GHG EMISSIONS AND ENERGY USE

Key points from the discussion are summarized here, as reported back to the group at large by Greg Keoleian, the Peter M. Wege Endowed Professor of Sustainable Systems at the University of Michigan. According to Keoleian, the group observed that the matrix (Table 6-1) could be useful, but decided that filling it out would have been too time consuming, given how much is already known about GHG emissions and energy use.

Methods

LCA was perceived as the tool of choice for evaluating both GHG emissions and energy use in a comprehensive way by individual participants of the group. With respect to GHG emissions, LCA could be used to evaluate both major emissions (i.e., CO_2, CH_4, N_2O) and minor emissions (e.g., chlorofluorocarbons used as refrigerants; perfluorocarbons used during aluminum production). Some emissions data are available at the national level (e.g., Environmental Protection Agency [EPA] reports on emissions associated with various agricultural activities, such as CH_4 emissions related to rice cultivation). With respect to energy use, LCA could be used to account for all of the various energy carriers across the total fuel cycle (e.g., transport fuels, electricity), as well as for upstream energy sources (e.g., feedstock for agricultural chemicals and fertilizers). With respect to which analytical approach to take, one could conduct either an attributional or consequential LCA, depending on the research question. (See the summary of Marty Heller's presentation in Chapter 3 for descriptions of the two approaches.)

Defining System Boundaries and Unit of Analysis

While LCA extends across all stages, from feed to end-of-life (i.e., feed, farm operations, processing, retail, consumption, disposal or end-of-life), the analysis could be truncated so that only certain components are evaluated. For example, in a comparison of agricultural production methods, it would not be necessary to include product packaging. According to Keoleian, the group spent a great deal of time discussing the importance of defining the functional unit, that is, the basis of analysis (e.g., kilograms [kg] of meat, kg of protein, calories, total nutrition or diet), and the importance of defining the temporal and spatial boundaries of the analysis.

Effects to Consider

Many GHG emission and energy use effects are quantifiable. The "real issue," Keoleian reported, is uncertainty. Many effects are difficult to accurately estimate. Keoleian described CO_2 from fuel combustion, N_2O from soils and manure, and CH_4 from manure and enteric fermentation, all of which contribute to climate change, as important direct costs to consider; and CO_2 sequestration resulting from certain types of land use changes (e.g., converting marginal land to rangeland) as a potential benefit to consider. Indirect land use change impacts can be important, but are difficult to quantify.

With respect to energy usage, Keoleian reported that British thermal units of primary energy consumption can be quantified "pretty well." LCA can also be used to quantify impacts from air pollutant emissions associated with energy usage (e.g., NO_x, PM, Hg, SO_2 from coal combustion); water pollution (e.g., nitrate run-off from corn production, oil spills); and land use impacts such as biodiversity loss (e.g., from surface mining of coal).

Measurement Challenges

Multiple working group participants emphasized several measurement challenges:

- *A high degree of uncertainty in characterizing non-CO_2 emissions*—Several different chamber and field measurement methods could be used to estimate the emissions. Using Intergovernmental Panel on Climate Change factors for ruminant enteric fermentation emissions, for example, is not an accurate method to estimate these emissions. According to Keoleian, many working group participants emphasized the development of new biogeochemical models that help to better characterize emissions.
- *Heterogeneity in production methods*—Different production methods can have different impacts, yet LCA data tend to offer limited resolution of these differences, calling for more extensive research exploiting the heterogeneity.
- *Non-GHG air pollutant emissions have both regional and local effects*—The impacts related to emissions of SO_2, NO_x, mercury, and other pollutants are more site dependent than GHG emissions, so it is important for the location of the emissions to be inventoried. Unfortunately, many databases do not report emissions in a spatially explicit manner. On the other hand, for carbon emissions and climate change impact, it does not matter where the greenhouse gases are released.
- *Allocation rules can influence results*—Allocation rules are used to distribute impacts from processes with coproducts across the various outputs (e.g., allocating feed production burdens to milk, butter, and hides).

Key Drivers

The working group participants discussed several activities that drive emissions and energy use: feed production, enteric fermentation, manure management, food storage (i.e., refrigeration), and food waste. For example, an estimated 26 percent of edible food is wasted. Participants highlighted several potential improvement strategies for countering these effects: adjusting animal rations and managing feed quality; harvesting energy from manure through anaerobic digestion; substituting renewable energy sources; and shopping more frequently to reduce household refrigeration.

Trade-Offs Related to Alternative Strategies

Many participants of the working group recognized that, when considering GHG emissions and energy usage, particularly when considering policies and interventions aimed at reducing GHG emissions or energy usage, one must also consider the human health, environmental health, and economic trade-offs. For example, with respect to the size of a production operation, while some concentrated animal feeding operations (CAFOs) may be more efficient than smaller operations, there may be trade-offs with respect to water quality, manure management, and other effects.

The LCA framework is very good at characterizing the effects of both production and consumption, particularly with respect to GHG emissions and energy usage. While the framework can be used to also characterize human health (e.g., via quality-adjusted life years) and other social impacts of production and consumption, Keoleian reported, "uncertainty increases tremendously when you start to look at some of these other effects." However, the LCA framework can be very useful in identifying "order of magnitude" trade-offs between health and environmental impacts. LCA can also be used to evaluate economic impacts of production and consumption, including both private and social costs (e.g., the "social cost of carbon," that is, monetized damages associated with increasing carbon emission) (Interagency Working Group on Social Cost of Carbon, 2010).

Research and Data Needs

Data needs depend on the question(s) being addressed. For example, if a goal is to characterize differences in production methods, then data would be needed for each type of production method (i.e., as opposed to industry average). With respect to data needs for specific stages of the food life cycle, many working group participants indicated there could be better data on the generation of food waste (e.g., data on spoiled milk is decades old) and better data on consumption patterns. Spatially explicit production data will also be necessary to capture impacts of categories that have spatially influenced characterization factors (e.g., water use, eutrophication, land use).

SOIL, WATER, AND OTHER ENVIRONMENTAL CONSEQUENCES OF FOOD PRODUCTION, PROCESSING, AND CONSUMPTION

Participants in this group spent most of their time discussing challenges to characterizing the soil, water, and other environmental consequences of the food system, as reported back to the group at large by Justin Derner, research leader for the Rangeland Resources Research Unit of the USDA Agricultural Research Service.

The Challenge of Heterogeneity

Working group participants discussed several major challenges to analyzing the external costs of animal production. One main challenge is the heterogeneity among sites with respect to practices, soils, climate, landscape, plant communities, and data (e.g., some sites have plentiful data, others none). Also, effects occur across variable spatial scales (e.g., small-scale farms versus large rangelands that encompass hundreds of thousands of hectares and may be publicly managed) and temporal scales (e.g., short-term versus long-term effects). On top of all this already existing heterogeneity, climate is not only changing, but it is changing differentially across the landscape, and the human population is growing, creating new food demands.

Building a Framework

It was suggested that one way to build a framework for addressing the environmental costs of the food system is to consider the threshold or cut-off rates of application beyond which four key elements—carbon (C), nitrogen (N), phosphorous (P), and sulfur (S)—become pollutants instead of nutrients. The analysis would very site-specific, but at least it would provide a framework for moving forward.

Effects to Consider

Although they did not identify externalities in the pure economic sense of the word, the group participants considered a wide range of effects: soil water erosion, soil wind erosion, soil fertility, water quality, water quantity, water scarcity, biodiversity, air quality/odors, pesticides, herbicides, open spaces, genetically modified organisms (plant and possibly animal), land use change, and deforestation. Additionally, there are several fairly well-known public health effects to consider in relation to some of these environmental effects, for example, asthma and mental health effects associated with exposure to certain odors. Several other considerations not captured in terms of monetization came up during conversation: quality of life; connection to the land; the value of open and green space; animal welfare issues; salt accumulation in soils; the value of wildlife habitat; ecosystem resilience (i.e., some ecosystems are resilient even after abuse, and show no change even when "pushed to the limits," while others are more fragile and undergo dramatic changes); and weed resistance to herbicides (e.g., some herbicides induce dramatic changes in ecosystem production).

Sources of Information and Challenges in Analyzing the Data

With respect to data, plentiful data are already available in various data networks and databases. Derner mentioned the Long-Term Ecological Research (LTER) Network and the new National Ecological Observatory Network (NEON), both funded by the National Science Foundation (NSF); the Long-Term Agro-Ecosystem Research (LTAR) network and Greenhouse gas Reduction through Agricultural Carbon Enhancement network

(GRACEnet), both coordinated by the U.S. Department of Agriculture's Agricultural Research Service (ARS); AmeriFlux; and citizen science efforts (i.e., public participation in scientific research). In addition to these data-collecting networks, several existing databases could be useful, such as the Conservation Effects Assessment Project (CEAP) database, long-term data at many sites, meta-analyses data, and remote sensing data.

While the data may be plentiful, so too are the limitations to analyzing those data. For example, the group struggled with identifying a benchmark for analysis. That is, what qualifies as "conventional" practice?[2] "Conventional" practices evolve over time. Additional challenges include the detection of "improper" management; the spatial distribution of manure/urine from animals; legacy effects of the dust bowl (e.g., huge soil losses); a realization that the global supply of phosphorous is limited and predicted to be depleted in less than a century, with consequences for cropping systems; the likelihood that there may be "sensitive areas" of high concern that could be targeted for sampling, with a cluster analysis focused on those areas; and water laws/rights and their impact on the cost of food.

CONSEQUENCES OF ANTIMICROBIAL USE IN AGRICULTURE

This group focused most of its discussion on swine production, reported facilitator Michael Doyle, Regents Professor of Food Microbiology and director of the Center for Food Safety at the University of Georgia.

State of the Evidence

Scientists have used a variety of tools—epidemiology, risk assessment, and molecular biology—to collect evidence on the public health impact of the use of antimicrobials in food production. Much of the epidemiological evidence resides with the U.S. Centers for Disease Control and Prevention (CDC). Most of the evidence is indirect findings related to antimicrobial use in agriculture.

Key challenges to collecting even indirect evidence, but especially direct evidence, stem from the complexity of the emergence of antibiotic resistance. This resistance can emerge in any of several ways: transfer between species, acquisition from the environment, selection, or co-selection. Co-selection occurs when use of an antibiotic selects not only for a resistance gene against the antibiotic being used, but also for resistance genes against other antibiotics. "We need to learn considerably more about co-selection," Doyle reported.

When thinking about the public health impact of antibiotic resistance,

[2] Some workshop attendees disapproved of the use of the word "conventional."

it is important to consider not just increased morbidity and mortality, including the potential for untreatable disease (e.g., systemic *Salmonella* infection that would be untreatable with antibiotics), but also the fact that antibiotic use in food production creates an environmental reservoir of antibiotic-resistant genes that includes non-pathogenic bacteria (i.e., nonpathogenic bacteria can harbor antibiotic-resistant genes that can be transferred to pathogenic bacteria).

Measurement Challenges

There was disagreement about the degree of evidence needed to establish a relationship between the use of antimicrobials in animal food production and human health and the feasibility of obtaining risk assessment data. Some individuals suggested that risk assessments are necessary and have been used successfully in the past to address this issue. Others expressed concern that risk assessments modified to evaluate risk from the use of antimicrobials in agriculture would be too costly and that conducting risk assessments on every antibiotic in every animal species would not be feasible.

Trade-Offs Related to Alternative Strategies

Working group participants highlighted several trade-offs that could be considered when evaluating the effects of antibiotic use in food production. First is productivity, with antibiotic use resulting in a more rapid growth rate and increased productivity, which in turn can reduce production costs (e.g., a more rapid growth rate can result in less manure and thereby lower the cost associated with removing manure). A second trade-off to consider is animal health and welfare. Third is food safety, with slower growth rates sometimes being associated with increased prevalence of disease. In high-intensity poultry production (i.e., with the use of antibiotics), the average time to grow a chicken from 1 day to age of processing is 42 days. Without the use of antibiotics, the average time increases by several days. The longer the production time, the greater the risk of *Campylobacter* colonization of poultry, a major cause of foodborne diarrheal illness in humans. A fourth trade-off to consider is profit, with discontinuation of the use of antibiotics being an increased cost for the farmer in animal growth rates and increased potential for disease. A final trade-off to consider is that a more efficient production process creates less waste per pound of product.

Research and Data Needs

Multiple working group participants voiced support for several research and data needs that would improve understanding of the public health effects of antimicrobial use in food production. Participants suggested more research could be conducted specifically to meet the following data needs:

- Data on the use of antimicrobials in agriculture—not just which antimicrobials are being used for which animal species, but also how they are being used with respect to dose, duration, and frequency
- Data on antibiotic use in humans (same types of data as listed above)
- Data on the evolution and transfer of resistance genes in different types of bacteria
- Data on co-selection
- Data on the prevalence of antimicrobial-resistant microbes—that is, the number of animals in an animal production facility actually carrying resistant strains of potentially harmful microbes
- Data on the impacts of different farm practices on disease management (e.g., Doyle suggested that Denmark would be a good place to start with respect to studying the impact of different farm practices, given its major strides in reducing antimicrobial use without impacting production cost or efficiency)

Many participants also emphasized possible improvements to the National Antimicrobial Resistance Monitoring System, which is the main system used to monitor antimicrobial resistance in animals, humans, and meats. While the system has been up and running for about 10 years and has revealed some trends, there are concerns that it is not well integrated (i.e., the database could be redesigned in a way that makes it easier to correlate antimicrobial resistance trends in animals, humans, and meats), and that it does not monitor emergent pathogens not traditionally found in foods (e.g., methicillin-resistant *Staphylococcus aureus*, *Clostridium difficile*, urinary tract *E. coli*).

PUBLIC HEALTH EFFECTS

Participants in this group worked with the matrix (Table 6-1), not always exhaustively filling in the matrix worksheet, rather discussing the six key items in turn. Included here is a summary of the report-back to the group at large by Sandra Hoffmann, senior economist with the Food Economics Division of the USDA Economic Research Service. The report identified 10 broad lessons drawn from the working group's discussion.

Potential Public Health Effects

Many participants in this group recognized that the American food system provides significant health benefits, in particular the provision of affordable nutrition, but given the task of the workshop, discussion focused on adverse health effects. Working group participants discussed a wide range of potential adverse health impacts from food production, processing, marketing, and consumption. Among these were

- acute and chronic illness from foodborne pathogens and parasites (e.g., enterohemorrhagic *E. coli* in beef, *Salmonella* in poultry or produce, and parasites like *Toxoplasma* gondii);
- the effects of exposure to chemicals (i.e., drug residues, hormones, and environmental toxins);
- diet-related chronic disease (e.g., diabetes, cardiovascular disease, cancer);
- occupational injuries and disease associated with agricultural production and food processing;
- adverse health effects associated with transportation (e.g., motor vehicle crashes, effects of air pollution);
- effects of exposure to air and water pollution from production practices (e.g., pesticide drift, manure-related ammonia emissions, and polluted surface water);
- mental health impacts (e.g., mental stress associated with living or working near concentrated animal feeding operations [CAFOs] or with living and working conditions among migrant laborers); and
- social impacts (e.g., effects of CAFOs on independence of rural communities, rural development, ability to conduct social or leisure activities) (see also Donham et al., 2007).

There was discussion about how much evidence of causality, as opposed to association, is necessary to identify an effect. The working group participants viewed their task for this exercise as discussing the scope of possible adverse health effects. Many participants recognized the importance of further work that would help to establish causality and to quantify the extent of the impacts. As one participant said, "This is just hypothesis generation at this point."

Measurement/Limitations

Working group participants discussed the availability and usefulness of different data sources that could be used to quantify these impacts. In general, there are limits to the usefulness of disease surveillance data in providing a comprehensive picture of health patterns associated with food production and consumption. Chemical exposures in agricultural production can result in acute illness. Those poisoned may seek care and cases may be reported to public health authorities. But there can be long latency periods between chemical exposure and illness, making it difficult to establish causation through surveillance data. Dietary exposure is typically very low-level, though potentially over long time periods. For these reasons, disease associated with chemical exposure is typically based on estimates of exposure and dose-response rather than surveillance data. Surveillance data are more useful in quantifying foodborne illness from pathogens or parasites, which are frequently associated with an acute onset of symptoms. But even then, most who suffer from these acute illnesses do not seek medical care and care providers may not report illnesses they do see. In addition, the availability of medical care varies geographically and by socioeconomic group. Active surveillance is not conducted on many of the outcomes of concern or in many areas of the country. As a broad generalization, the quality of data on exposure and dose-response modeling is generally stronger for chemicals than for foodborne pathogens and parasites, and the quality and availability of disease surveillance data are generally stronger for pathogens and parasites than for chemical hazards. Federal agencies, state departments of public health, and possibly some private-sector organizations will be important sources of data; examples include the EPA for air and water pollution exposure data, the National Institute for Occupational Safety and Health, state health departments and possibly labor unions for occupational safety data, and the CDC for surveillance data on foodborne illness and other health outcomes. The 2003 Institute of Medicine (IOM) report *Dioxin and Dioxin-Like Compounds in the Food Supply: Strategies to Decrease Exposure* provides data on exposure to dioxin in food (IOM, 2003).

Externality or Not?

Group participants discussed which effects could be considered externalities. There was lively debate about whether a full-scale accounting of the food system should even cover external effects that have been internalized. For example, participants debated the extent to which the cost of diet-related cardiovascular disease is internalized through health and life insurance premiums and therefore not an externality and should not be included in the analysis. Many participants agreed that even if the cost of health care were fully internalized through insurance premiums, the impact of disease extends beyond the ill person and their immediate family, and therefore social costs likely exceed medical costs. As a result, the full social cost is not internalized by individuals, and would be important to reflect in external costs. Group participants recognized that the analysis required to adequately address these kinds of questions was beyond this scoping exercise. In the spirit of viewing the discussion as an effort in hypothesis generation, participants decided to identify all potential health effects, regardless of whether those health effects qualify as externalities.

Trade-Offs Related to Alternative Strategies

The group had limited time to discuss trade-offs among alternative strategies to reduce external costs associated with food production and consumption. One set of trade-offs discussed was those related to large- versus small-scale production. Potential benefits of CAFOs include an economy of scale that affords more efficient sewage and manure management and, in some cases, improved control of some pathogens. For example, trichinosis from pork has been significantly reduced by the improved rodent control made possible by confined feeding operations. Potential costs include the mental health and community effects where CAFOs are located, and possibly greater prevalence of other pathogens or greater use of antibiotics among CAFOs compared with smaller-scale livestock operations.

Lessons Learned

Hoffmann summarized what she viewed as the 10 major lessons from the working group exercise:

1. The matrix did not include some important dimensions of the problem components. In particular, it did not provide a place to include the magnitude of the impact and confidence about the magnitude of impact. It also did not provide a place to note the distribution of impacts. For example, impacts may vary by geographical location or by income, age, or social groups.
2. The concept of externality might not be the best way to frame the analysis because it does not capture or allow expression of some major concerns in the public health community. For example, cardiovascular disease is a major cause of death in the United States. Yet, it is not clear that diet-related cardiovascular disease is an externality.
3. Many participants of the working group felt that more consideration could be given to methods or approaches for capturing the social and individual impacts of large-scale production, for example, impacts of CAFOs on local social networks and local energy use.
4. Focusing exclusively on adverse health impacts of food production and consumption without also looking at the health benefits may provide a distorted picture. For example, while excessive red-meat consumption can contribute to cardiovascular disease, meats also provide a high-quality source of protein. There was also some discussion on whether an examination of nutritional benefits should focus on individual foods or dietary patterns at large.

5. The industrial structure of production is important to take into account. For example, the cost of seeking other work with fewer occupational safety risks is greater in rural areas where meat processing plants are isolated and where workers have fewer options for employment.
6. Regional concentration is also important to consider. For example, toxic algal blooms have appeared in areas where there is high regionally concentrated agricultural production, such as in the Delmarva Peninsula in the Chesapeake Bay (where poultry production is concentrated) and in the Carolinas (where hog production is concentrated).
7. Capturing all stages of the life cycle, not just production and consumption, would be useful. For example, the group did not discuss food preparation by the retail sector of the food industry. Yet, the public health impact associated with the addition of salts, nitrates, or other additives to foods in restaurants or other retail establishments could be considered an externality if consumers are unaware of the addition or risks of those substances.
8. Considering production methods is important when evaluating health impact. Certain foodborne illnesses are reemerging in association with changes in management practices. For example, trichinosis is reemerging in association with field-raised hogs, and some dairy-related illnesses are reemerging with a loosening of norms around pasteurization.
9. While the group discussion was a great brainstorming exercise, moving forward will require a very solid literature review and analysis of available data. The matrix helped facilitate group discussion, but it may not be the best structure for more in-depth analysis.
10. It is important to define the scope of the effects to be considered in a full report. For example, is the goal to examine only direct public health effects of food consumption or to more broadly examine indirect effects of the production process as well (e.g., occupational illness)? Also, is the goal to examine acute effects, chronic effects, or transgenerational effects?

MAJOR OVERARCHING THEMES OF WORKING GROUP DISCUSSIONS

Although the groups took different approaches, based on the report-backs to the group at large, the group discussions shared several major overarching themes:

- Many public health and environmental costs can be quantified, but there is a great deal of uncertainty about many estimates. For example, with respect to energy and emissions, most non-GHG emission amounts are significantly uncertain.
- Production systems are highly variable, not just with respect to methods (e.g., large- versus small-scale production), but also site specificity (e.g., local soils, climate, landscape), which has implications not just for analysis but also for data collection.
- Even people with very different perspectives together voiced their support for more information that would allow for improved decision making about many of the issues. For example, many data sources could be updated, particularly with respect to consumption.
- Many questions about the scope of effects need to be considered, with varying opinions about whether the concept of externality is the best way to frame a full-scale accounting of the "true costs" of food. All four break-out groups struggled to understand exactly what to measure—externalities as defined by economists or all external effects regardless of whether they qualify as externalities (e.g., external effects that are internalized).
- All groups recognized the importance of trade-offs. Indeed, workshop chair Helen Jensen began the break-out group report-back session by commenting on the April 2012 announcement that the European Union would be banning sow stalls beginning in January 2013. The predicted 5-10 percent price increase for pork as a result of the ban is a good example of the type of trade-off that needs to be considered when evaluating the effects of different regulations or practices (European Commission, 2012).
- Jensen observed that everyone began to get a better sense of the food system and began to see problems somewhat differently during the small group discussions.
- The groups were largely brain-storming exercises. There were several calls for a more systematic approach to identifying effects and methodologies for measuring those effects. For example, one participant suggested that a systematic survey of the literature would yield more comprehensive lists of effects, trade-offs, methodologies, and limitations of those methodologies.

REFERENCES

Donham, K. J., S. Wing, D. Osterberg, J. L. Flora, C. Hodne, K. M. Thu, and P. S. Thorne. 2007. Community health and socioeconomic issues surrounding concentrated animal feeding operations. *Environmental Health Perspectives* 115(2):317-320.

European Comission. 2012. *Animal welfare: Commission steps up pressure on member states to implement ban on individual sow stalls.* Press release. http://europa.eu/rapid/pressReleasesAction.do?reference=IP/12/404 (accessed October 19, 2012).

Interagency Working Group on Social Cost of Carbon. 2010. *Technical Support Document: Social cost of carbon for regulatory impact analysis, Under Executive Order 12866.* Washington, DC: Environmental Protection Agency. http://www.epa.gov/oms/climate/regulations/scc-tsd.pdf (accessed November 5, 2012).

IOM (Institute of Medicine). 2003. *Dioxins and dioxin-like compounds in the food supply: Strategies to reduce exposure.* Washington, DC: The National Academies Press.

7

Reflecting on the Path Forward

Throughout the workshop, individual participants reflected on the presentations and discussions and shared key considerations for strategies going forward. This chapter is a compilation summary of all the open discussions, organized into five major themes:

1. The concept of externality and whether externalities are the best way to frame a full-scale accounting of the cost of food
2. The importance of recognizing trade-offs among costs and benefits, for example, the trade-offs associated with large- versus small-scale production
3. The challenge of quantifying effects and the level of uncertainty around many estimates
4. Opportunities for more data and research, but also concerns that sufficient data in some areas are being overlooked
5. The daunting challenge of measuring a single "true cost" of food

ARE EXTERNALITIES THE BEST WAY TO FRAME THE PROBLEM?

Participants expressed some disagreement about whether certain effects should be included in a full-scale accounting of the cost of food. Some participants argued that even though the external cost of diet-related cardiovascular disease may be internalized through health insurance premiums, cardiovascular disease nonetheless imposes a cost to society that exceeds, or is different than, its internalized value. For example, during his presentation, Steven Wing argued that many external costs of the food system are strongly related to health inequalities, with certain populations bearing a disproportionate amount of those costs. That unequal distribution is not accounted for in a model based on externalities. Some participants wondered if there is another model or approach that could be used that accounts for distribution. The implications of building a framework based on a broader view of costs are unclear.

James Hammitt offered an explanation. The concept of externality falls under the purview of welfare economics, which is an attempt to quantify the well-being of people. When quantifying the costs and benefits associated with an activity, welfare economists quantify in dollar amounts those costs and benefits as perceived by the individuals actually benefiting or being harmed and then add those quantities across society. If some policy action is expected to have society-level benefits that exceed society-level costs, in principle those people who gain from the policy could "transfer" some of their gains to those who are harmed so that everyone would be better off after the transfer is made. In other words, Hammitt said, "you can think of it as expanding the size of the social pie." If externalities can be corrected through policy, that is, if they can be internalized, then "we can make the pie bigger." But how that pie is divided across people is a separate question. He said, "If we don't like the distributional effects of that, we can redistribute using other mechanisms. Rather than foregoing the opportunity to increase the size of the social pie, we should go ahead and increase it, and then redistribute it." Not implementing a policy that imposes costs on a subset of the subpopulation while providing benefits that exceed costs overall is, in his opinion, an "extreme position." There are more efficient ways to redistribute well-being than to not implement a policy because it will increase the cost to a subset of the population. In the case of a food policy that increases overall well-being but increases the cost of food to poor people, he said, "It is much better to deal with the poverty directly."

Anna Alberini added that, although economists are primarily concerned with the size of the pie, few agencies conduct cost-benefit analyses without also conducting regulatory impact analyses to deal with those distributional issues. She said, "Considerations of this kind do indeed take place."

Still, some participants wondered whether there might be other economic strategies, such as ecological economic models, that could provide other ways to frame the discussion. Helen Jensen responded that, yes, there are other approaches, but none of those approaches take away from the basic understanding of externality. Jayson Lusk warned, "If we are not going to use the externality argument, there needs to be some rational argument for what the basis is of some policy recommendation." He explained that externalities represent an opportunity for interventions that allow people to benefit by their own account. The alternative is paternalism. He said, "With children, most people are more open to using paternalism as a justification for public policy. People like myself are a little less willing to use that justification on the average population."

Lusk's response triggered some comments about choice and how behavior in the real world differs from behavior in economic models. For example, Aaron Wernham expressed concern that the policy environment might drive choice as much as the market does. Also, earlier during the workshop, there had been a question about whether lack of choice in food deserts (i.e., areas without access to food) can be factored into a model based on externalities and responding remarks about how lack of access to food is a social issue, not a market issue, and therefore cannot be analyzed within the context of externalities. (See the summary of James Hammitt's presentation on public health effects in Chapter 3 for more thoughts on the challenge of analyzing effects impacted by "real-world" behavior.)

TRADE-OFFS[1] ASSOCIATED WITH DIFFERENT SCALES OF ANIMAL PRODUCTION

The importance of trade-offs was a major overarching theme of the workshop discussion, from keynote speaker Katherine Smith's admonition that "everything is relative" onward. Smith cautioned that a cost is only a cost relative to something else. Most subsequent discussion of trade-offs revolved around those associated with large- versus small-scale production. An audience member remarked that agriculture in the United States has undergone a major transformation over the past century from many small independent producers to a small number of large, sometimes global, corporations controlling sectors of the food system. At the same time, according to Michael Doyle, there has also been a recent drive in the United States and across Europe toward a small-scale way of farming food animals. These trends raise the questions: What are the costs and benefits of large- versus small-scale animal production? What are the implications for food safety, water and air quality, and manure management? How are local communities impacted? What are the costs and benefits to the farmers themselves?

In some experts' opinion, arguably one of the greatest benefits of large-scale production is its economy of scale, with large-scale production being more efficient and more cost-efficient. However, one audience member remarked that the economy of scale afforded by large-scale production may shift at a certain "tipping point," beyond which further production actually creates greater costs than benefits. For example, there may be a point at which raising too many pigs turns manure from a commodity into

[1] A trade-off is defined here as an exchange of one effect for another when a different decision, policy, or practice is implemented.

a toxic problem. The participant wondered how that "tipping point" could be factored into a true-cost accounting of the food system.

A challenge brought up in regard to meat production was how to design a different system that would produce the amount of meat equivalent to that produced by concentrated operations and in a safe manner. Presumably, the costs to the environment of a small animal production system versus a concentrated production system would be similar as long as the total amount of animals produced is the same. In this respect, a participant remarked one of the goals should be to raise fewer total animals. He said, "The epidemiological evidence is overwhelming that our high meat diet is unhealthy."

John Antle identified loss of farmer income as a cost of small-scale animal production. Much of what is driving the trend toward large-scale farming, he said, is the desire to generate household incomes that are comparable to those earned by professionals in nonagricultural sectors. For example, a wheat farmer in Montana cannot generate an income greater than about \$30,000-\$40,000 a year without more than 3,000-4,000 acres of land. He said, "you have got to keep those factors in mind. . . . There are fundamental economics driving what we see in terms of the scale of production." Some audience members agreed that economics are driving the trend toward large-scale production, that any policy changes aimed at reducing some of the external costs associated with large-scale production would need to be done very carefully, and that choices about trade-offs would be paramount.

An audience member observed that the export of finished meat products is another factor driving the trend toward large-scale production, and that the cost of global trade also needs to be considered when evaluating the trade-offs associated with small- versus large-scale production.

UNCERTAINTY ABOUT THE MAGNITUDE OF SOME EFFECTS

Another major overarching theme of the open discussions was the challenge of quantifying effects. Throughout the workshop, participants considered a range of methodologies for quantifying health, environmental, and other impacts. Some methodologies seem especially well suited for certain effects. For example, participants in one of the working groups described LCA as the tool of choice for examining GHG emissions. But for other effects, like the cost of antimicrobial use in food animals, varying opinions were expressed on whether risk assessment would be a feasible strategy for covering all antibiotics across every animal species. Some workshop participants opined that the greater challenge is not quantifying effects, rather it is quantifying them with certainty. Many participants noted that the level of certainty about the magnitude of effect varies tremendously, with some effects being very clearly associated with sources and others not. Indeed, some effects cannot be quantified at all. Hammitt mentioned that costs in the NRC (2010) energy report were calculated with uncertainty ranges (see Table 5-1 in Chapter 5).

Several factors contribute to this uncertainty, not the least of which is the heterogeneous nature of the food landscape. Several participants commented at numerous times on the tremendous variation in production that exists across both space and time, which creates both analytical and data challenges. Another contributing factor is the complexity of the food system and the challenge of teasing apart pathways. As just one example, while almost the entire workshop discussion was on food products, an audience member pointed out that the same systems that produce animal food products also produce nonfood by-products (e.g., hides, fats, pharmaceutical products). He asked how costs and benefits should be allocated between food products versus nonfood by-products. Hammitt replied that there is "no non-arbitrary way" to allocate external costs (and benefits) associated with raising animals across all of the various products. Rather than considering total costs and how to allocate those costs, he suggested considering marginal changes that would occur if animals were raised in a different way. John Antle opined that if externalities were addressed at the production level (e.g., by taxing or otherwise imposing measures that impact industry decisions about production), they would be appropriately valued into the system at that level and reflected in the cost of product and by-product production. Marty Heller agreed with Antle that the problem could be approached from that sort of "system expansion perspective," whereby all of those other components (i.e., by-product production pathways) are incorporated into the analytical model, but cautioned that an analysis of impacts of the food system in particular may still require some sort of allocation of costs among food versus nonfood products.

OPPORTUNITIES FOR MORE DATA AND RESEARCH

For some effects, lack of sufficient data may also contribute to uncertainty about the magnitude of those effects. There were many calls throughout the workshop for more data and research. For example, there was a call during the public health effects break-out group for more research on the health effects of exposure to hormones in animal food products. As another example, during her presentation Anna Alberini listed four understudied areas that she thinks represent fertile ground for new research on food-related valuation: (1) willingness to pay (WTP) to reduce dietary cardiovascular and diabetes risks; (2) WTP to reduce endocrine disruption risks (i.e., which have been linked to certain pesticides); (3) WTP to reduce effects on the reproductive system; and (4) WTP to reduce antibiotic resistance.

For some phenomena, it is not clear whether more data are needed or whether existing data need to be more thoroughly analyzed or interpreted. For example, one participant asked whether more research is needed on populations that tend to be excluded from research studies. Steven Wing responded that, in some cases, "we have a lot of data, but we are not paying attention to the data that we have." The greater challenge, in his opinion, is the question(s) being asked. He mentioned a long history of research on questions that are of economic interest to producers and a short history of research on questions related to the health and environmental impacts of production systems. "So do we need more [data]? Maybe we do," he said. But the lack of information is also a result of "who is at the table." As another example, a participant questioned the call for more scientific evidence on the association between antimicrobial use in food animals and antimicrobial resistance in humans. He implored that scientists have known about the cost of antimicrobial resistance for decades, since the 1969 Swann Committee report (Swann et al., 1969). Yet, the demand for more data persists. Why? He called for more discussion on the political nature of the debate about antimicrobial use in food animals.

THE DAUNTING CHALLENGE OF MEASURING "THE" COST OF FOOD

Early on during the workshop, an audience member commented on the complexity of the food system and its wide range of effects and asked whether there was a way to ensure that all costs and benefits have actually been measured. He said, "What if I miss something? . . . The ultimate answer would be just wrong." Even if the focus is on marginal costs, not total costs, still the dimensions of that margin need to be known. "To some extent," he said, "I have sort of despaired listening to this conversation." John Antle expressed similar concern about the wide range of effects, noting that policies that fail to consider important consequences "really mess things up."

Given what she characterized as "squishiness" from a lack of data and problems with analyzing those data, Katherine Smith questioned the intention of tallying up all costs and benefits to derive an estimate of the total "true" cost of food. She suggested evaluating the effect of public policy on one "dimension" or on the trade-offs between a couple of dimensions of the food system, instead of calculating total cost. Jayson Lusk and others agreed that analyzing the costs and benefits of specific interventions might be a more feasible research strategy than estimating the "true" cost of food.

WRAP-UP

There is no obvious best research strategy for conducting a full-scale accounting of the external costs and benefits of the food system. Activities at all stages of the food life cycle have tremendously far-reaching and wide-ranging consequences for the environment, human health, the economy, and society at large. Workshop participants considered a range of methodologies to consider, from LCA to HIA, yet many questions remain about how to quantify effects with an acceptable level of certainty and how to analyze effects that cannot be quantified. Measuring and valuing the "true" cost(s) of food is made all the more difficult by widely divergent expert opinions about how to even frame the challenge—within the context of economic thinking about externalities, or otherwise. The participants had differences of opinion on whether focusing on externalities in the strict economic sense (see Box 1-1) is too limiting.

While the goal of the workshop was not to reach any conclusions or make any recommendations, some personal opinions were expressed about how to move forward. There were a couple of calls for conducting a more systematic and comprehensive consideration of potential effects, methodologies for measuring those effects, and limitations of the methodologies. While the small group break-out session was a valuable exploratory exercise, the opportunity for even more work persists. For example, one group opted not to work with the matrix because it would have been too time-consuming given how much is known about GHG emissions and energy use. Another group did not have enough expertise at the table to consider some issues.

In addition to more thoroughly considering potential effects and methodologies for quantifying and valuing those effects, there were many calls for a reconsideration of the intention of a full-scale accounting of the "true" cost of food. Several participants questioned not just the feasibility, but also the applicability, of assembling a list, or matrix, of all potential costs and benefits and trade-offs, and suggested instead a more selective examination of the food system from a policy perspective. That is, examine a policy or intervention and its potential impact on costs and benefits rather than the costs and benefits of the food system as it is.

While a full-scale accounting of the external costs and benefits of the food system will undoubtedly be a challenging endeavor, this information-gathering workshop was an important first step in showcasing the range of expertise, methodologies, and information sources that could be used to pursue that endeavor. Although the U.S. food system provides multiple benefits, those benefits could be expanded even further with a better understanding of how decisions made along the entire course of the food life cycle adversely impact the environment, public health, and community economic development.

REFERENCES

NRC (National Research Council). 2010. *Hidden costs of energy: Unpriced consequences of energy production and use.* Washington, DC: The National Academies Press.

Swann, M. M., et al. 1969. *Report of the Joint Committee on the Use of Antibiotics in Animal Husbandry and Veterinary Medicine.* London: Her Majesty's Stationery Office.

A

Workshop Agenda

Exploring the True Costs of Food

April 23 and 24, 2012

The Pew Charitable Trusts
901 E Street, NW
Americas Room, Second Floor
Washington, DC 20004

Meeting Goals

- Discuss the environmental and public health effects and trade-offs of the practices that occur at all life cycle stages (e.g., production, processing, packaging, distribution, preparation, and consumption) for all foods in the U.S. food system.
- Identify the types of information sources and methodologies required to recognize and estimate the costs and benefits of environmental and public health consequences associated with the U.S. food system.
- Discuss potential issues and challenges to estimating/quantifying the hidden costs of the U.S. food system.
- Consider the kind of research strategy and feasibility of conducting a full-scale accounting of the environmental and public health effects for all food products of the U.S. food system.

DAY 1: April 23, 2012

8:00 a.m.	Registration
8:30	Welcoming Remarks *Helen Jensen, Workshop Planning Committee Chair* *Iowa State University*
8:35	Sponsor Remarks *Anne Haddix, National Center for Chronic Disease Prevention and Health Promotion, Centers for Disease Control and Prevention*
8:50	KEYNOTE The Economics of Food Prices and Considerations for Valuing Food *Katherine (Kitty) Smith, American Farmland Trust*
9:15	Q&A

Session 1 – Understanding Measures and Strategies for Estimating the Costs of Food

9:30	Life Cycle Assessment *Martin Heller, University of Michigan*
10:00	Health Impact Assessment *Jonathan Fielding, Los Angeles County Department of Public Health (via phone)*
10:30	*Break*
10:45	Environmental Consequences *John Antle, Oregon State University*
11:15	Public Health Consequences *James Hammitt, Harvard University*
11:45	Discussion
12:15 p.m.	*Lunch*

Session 2 – Identifying External Effects

1:15	Working Group Introductions *Helen Jensen*
1:30	Working Groups *(two rotations: 1:30-3:00 and 3:00-4:30)* • Energy and greenhouse gas emissions

- Soil, water, and other environmental consequences
- Consequences of antimicrobial use in agriculture
- Other public health consequences

4:30 Reflections and Reactions
All Participants

5:00 Adjourn

DAY 2: April 24, 2012

8:00 a.m. Registration

8:30 Welcoming Remarks
Helen Jensen, Planning Committee Chair

8:45 Reports from Working Groups

9:45 Panel on the Social and Ecological Dimensions of the Food Supply
Ecological services: *Scott Swinton, Michigan State University*
Health inequalities: *Steven Wing, University of North Carolina*
Accessibility to food: *Ricardo Salvador, Union of Concerned Scientists*
Animal welfare: *Jayson Lusk, Oklahoma State University*

Session 3 – Quantification Methods

11:00 Lessons from *The Hidden Costs of Energy* Report
James Hammitt, Harvard University

11:30 Valuing Agricultural Externalities and Public Health Impacts
Anna Alberini, University of Maryland

12:00 p.m. Concluding Thoughts and Discussion of Next Steps
Helen Jensen

12:15 Adjourn

B

Speaker Biographical Sketches

Anna Alberini, Ph.D., is associate professor of economics in the Department of Agricultural and Resource Economics at the University of Maryland, College Park. Her research interests are in environmental economics (including valuation of natural and nonmarket resources, estimation and valuation of health effects of environmental quality), energy economics, econometrics and statistics. Alberini has served as a co-editor of the *Journal of Environmental Economics and Management*, is serving on the editorial board of numerous environmental economics journals, and has served on the Science Advisory Board of the Environmental Protection Agency for Environmental Economics for two terms. She has participated in a number of research projects funded by the European Commission, and has done research for U.S. and Canadian government agencies. Dr. Alberini earned her Ph.D. in economics from the University of California, San Diego.

John M. Antle, Ph.D., is professor in the Department of Agricultural and Resource Economics at Oregon State University and a University Fellow at Resources for the Future. Dr. Antle previously served as professor at the University of California, Davis, and Montana State University. He was a senior staff economist for the President's Council of Economic Advisers and served as a member of the National Research Council's (NRC's) Board on Agriculture and Natural Resources. Dr. Antle is a fellow and past president of the American Agricultural Economics Association. His current research focuses on the sustainability of agricultural systems in industrialized and developing countries, including climate change impacts, adaptation, and mitigation in agriculture; assessment of environmental and social impacts of agricultural technologies; and geologic carbon sequestration. He received his Ph.D. in economics from the University of Chicago.

Justin Derner, Ph.D., is research leader for the Rangeland Resources Research Unit of the U.S. Department of Agriculture (USDA) Agricultural Research Service. Currently, Dr. Derner leads a multidisciplinary team of scientists developing and providing land managers with the necessary tools to address the interface of contemporary production-conservation issues related to provision of ecosystem goods and services on western U.S. rangelands. His research ascertains the effects of livestock as ecosystem engineers, alone or in combination with fire and prairie dogs, to influence vegetation heterogeneity, modify states of vegetation, and affect resilience within ecological sites of semiarid rangelands. Research efforts target management strategies for mitigation and adaptation of climate change on rangelands by evaluating dynamics of soil carbon and nitrogen as influenced by management X environment (weather/climate) effects. Dr. Derner is a principal investigator for the Central Plains Experimental Range site of the Long-Term Agro-ecosystem Research network, a co–principal investigator on the National Science Foundation- (NSF-) funded Shortgrass Steppe Long-Term Ecological Research Project. In addition, he is an affiliate faculty member in the Department of Ecosystem Science and Management at the University of Wyoming and the Department of Forest and Rangeland Stewardship at Colorado State University. He received his Ph.D. in rangeland ecology and management from Texas A&M University.

Michael P. Doyle, Ph.D., is a Regents Professor of Food Microbiology and director of the Center for Food Safety at the University of Georgia. He is an active researcher in the area of food safety and security and works closely with the food industry, government agencies, and consumer groups on issues related to the microbiological safety of foods. He serves on food safety committees of many scientific organizations and has been a scientific advisor to many groups, including the World Health Organization, Institute of Medicine (IOM), National Academy of Sciences (NAS)-NRC, International Life Sciences Institute-North America, Food and Drug Administration, USDA, Department of Defense, and Environmental Protection Agency. He is a Fellow of the American Academy of Microbiology, American Association for the Advancement of Science, and International Association for Food Protection and the Institute of Food Technologists, and is a member of the IOM. Dr. Doyle received his B.S., M.S., and Ph.D. degrees from the University of Wisconsin in bacteriology/food microbiology.

Jonathan E. Fielding, M.D., M.P.H., M.B.A., is the director of the Los Angeles County Department of Public Health, and a professor at the Schools of Public Health and Medicine at the University of California, Los Angeles.

He previously served as co-director for the Center for Health Enhancement, Education and Research. He also chairs the Secretary of Health and Human Services's expert advisory group on the 2020 Healthy People Project and the U.S. Community Preventive Services Task Force and is editor of the *Annual Review of Public Health*. His current research interests are health impact assessment and forecasting future health. He received his M.D. and M.P.H. from Harvard University and M.B.A. from the Wharton School of Business at the University of Pennsylvania.

Anne C. Haddix, Ph.D., is the senior policy advisor at the Centers for Disease Control and Prevention's (CDC's) National Center for Chronic Disease Prevention and Health Promotion. Dr. Haddix co-founded and successfully cultivated Prevention Effectiveness as a scientific discipline at CDC, establishing the first set of methodological guidelines for cost-effectiveness analysis of public health interventions. She also helped to create the Prevention Effectiveness Postdoctoral Fellowship program. Dr. Haddix is the author or co-author of numerous scientific publications, and editor of *Prevention Effectiveness: A Guide to Decision Analysis and Economic Evaluation*. Dr. Haddix received her Ph.D. in agricultural economics from the University of Georgia and her M.S. in agricultural economics and B.A. in biology from California State University, Fresno.

James K. Hammitt, Ph.D., is professor of economics and decision sciences at the Harvard School of Public Health and a visiting professor at the Toulouse School of Economics in France. Dr. Hammitt's research and teaching concern the development of decision analysis, cost-benefit analysis, game theory, and other quantitative methods and their application to health and environmental policy in the United States and internationally. His research includes work on global climate change, the risks of pesticides and other contaminants in food, and the cost-effectiveness of air pollution control strategies. He also studies ways to measure the value of reducing health risks, including monetary and health-adjusted life-year metrics. Dr. Hammitt previously served as a senior mathematician at the RAND Corporation and as the Pierre-de-Fermat Chair of the Toulouse School of Economics. He received his Ph.D. in public policy from Harvard University.

Martin Heller, Ph.D., is a research specialist with the Center for Sustainable Systems at the University of Michigan. Dr. Heller has conducted life cycle assessment studies of short-rotation woody biomass energy crops; a large-scale vertically integrated U.S. organic dairy (Aurora Organic Dairy); and as part of an international team, a comprehensive, spatially explicit study of U.S. dairy production for the Dairy Research Institute. He also developed a seminal report on life cycle–based sustainability indicators for assessment of the U.S. food system, published in *Agricultural*

Systems. Previously, as a researcher at the C.S. Mott Group for Sustainable Food Systems at Michigan State University, Dr. Heller investigated the ecological services provided by pasture- and confinement-based dairies, and developed a "community food profile" intended to frame for a general audience the opportunities of a community-based food system. Dr. Heller received a B.S. in chemical engineering from Michigan State University and a Ph.D., also in chemical engineering, from the University of Colorado at Boulder.

Sandra A. Hoffmann, Ph.D., is a senior economist with the Food Economics Division of the USDA Economic Research Service. Her research focuses on food safety, valuation of the health benefits of public policies, and integration of economic analysis and risk assessment. She is recognized for her research on the attribution of foodborne illness to its food sources and on childrens' environmental health. She has advised the Organisation for Economic Co-operation and Development (OECD) and the Environmental Protection Agency (EPA) on valuation of children's benefits from environmental health programs. Sandy was a research fellow at Resources for the Future (2000-2010) and a faculty member at the University of Wisconsin-Madison (1999-2000). She also practiced pesticide and chemical manufacture regulatory law (1986-1989) and served with the U.S. Peace Corps in rural Chile (1980-1982). Sandy earned her Ph.D. from the Department of Agricultural and Resource Economics, University of California, Berkeley, and her M.A. in agricultural economics from the University of Wisconsin–Madison. She also received her J.D. from the University of Michigan Law School.

Gregory A. Keoleian, Ph.D., is the Peter M. Wege Endowed Professor of Sustainable Systems at the University of Michigan with appointments in the School of Natural Resources and Environment and the Department of Civil and Environmental Engineering. Dr. Keoleian serves as director of the Center for Sustainable Systems. His research focuses on the development and application of life cycle models and sustainability metrics to guide the design and improvement of products and technology including energy systems, transportation, buildings and infrastructure, consumer products and packaging, and a variety of food systems. The center has pioneered the development of methods in life cycle modeling to evaluate the sustainability performance of food systems. Dr. Keoleian is serving a 2-year term as president of the International Society for Industrial Ecology. He received his M.S.E. and Ph.D. in chemical engineering from the University of Michigan.

Jayson Lusk, Ph.D., is a professor and Willard Sparks Endowed Chair in the Department of Agricultural Economics at Oklahoma State University.

He previously served on faculty at Purdue University and Mississippi State University. Dr. Lusk has published more than 115 articles in peer-reviewed scientific journals on topics related to economics, consumer behavior, and food marketing and policy. He serves on the editorial council for seven top academic journals, including the *American Journal of Agricultural Economics*, the *Journal of Environmental Economics and Management*, and the *Journal of Consumer Affairs*. He is a former director of the Agricultural and Applied Economics Association. Dr. Lusk recently coauthored a book on consumer research methods, an undergraduate textbook on agricultural marketing and price analysis, and a book on the economics of animal welfare. He also coedited the *Oxford Handbook of the Economics of Food Consumption and Policy*. His forthcoming book, *Food Police*, is slated for publication in 2013. He earned his B.S. in food technology from Texas Tech University and his Ph.D. in agricultural economics from Kansas State University.

Ricardo J. Salvador, Ph.D., is director and senior scientist in the Food and Environment Program at the Union of Concerned Scientists. Earlier, he served as associate professor at Iowa State University, where he was the charter chair of the graduate program in sustainable agriculture. His specializations range from drought resistance mechanisms in the maize crop to advanced crop production techniques and global food issues. Dr. Salvador previously served as a program officer for food, health, and well-being with the W.K. Kellogg Foundation, where he managed the Foundation's food system program. He earned his B.S. in general agriculture from the New Mexico State University and M.S. and Ph.D. in crop production from Iowa State University.

Katherine (Kitty) Smith, Ph.D., is vice president of programs and chief economist at the American Farmland Trust. She oversees research and policy development, and administers programs concerning farmland protection, food, agriculture, and the environment. Prior to joining American Farmland Trust, Dr. Smith served as administrator of the USDA's Economic Research Service (ERS). She has held numerous other leadership positions within the ERS, including director of the resource economics and market and trade economics divisions. Dr. Smith also served as the first director of policy studies, pioneering cutting-edge concepts such as "green payments" and whole-farm conservation planning with the Henry A. Wallace Institution for Alternative Agriculture. She has served on several United Nations Expert Panels and chaired the Organization of International Cooperation and Development's Joint Working Party on Agriculture and Environment. Her work has been published in books and scholarly journals throughout her career. Dr. Smith is a fellow of the Agricultural and Applied Economics

Association. She earned her B.S. in biological sciences and Ph.D. in agricultural and resource economics from the University of Maryland.

Scott M. Swinton, Ph.D., is a professor and associate chair in the Department of Agricultural, Food, and Resource Economics at Michigan State University. His current research explores economic approaches to enhance the provision of ecosystem services from agriculture, including projects with the NSF's long-term ecological research agroecological site in Michigan and the U.S. Department of Energy Great Lakes Bioenergy Research Center. In addition to his U.S. activities, he has conducted research on farming systems and natural resource management while living in Africa and Latin America. Dr. Swinton served on the NAS panel on the status of pollinators and on several journal editorial boards, including the *American Journal of Agricultural Economics* and *Frontiers in Ecology and the Environment.* He earned his B.A. in political science and economics from Swarthmore College, his M.S. in agricultural economics from Cornell University, and his Ph.D. in agricultural and applied economics from the University of Minnesota.

Steven Wing, Ph.D., is an associate professor at the University of North Carolina, Chapel Hill. His research and teaching are primarily in the areas of occupational and environmental health. Dr. Wing has conducted several studies of air pollution and health in communities near confined animal feeding operations. Dr. Wing received his Ph.D. in epidemiology from the University of North Carolina at Chapel Hill.

C

Workshop Attendees

Anna Alberini
University of Maryland

John Antle
Oregon State University

Michael Apley
Kansas State University

Peggy Barlett
Emory University

Sarah Bell
The 11th Hour Project

Jay Bhattacharya
Stanford University

John Blanton
The Samuel Roberts Noble Foundation

Dennis Boik
American Meat Institute

Betsy Booren
American Meat Institute

Roger Claassen
U.S. Department of Agriculture (USDA) Economic Research Service (ERS)

Caitlin Conover
Environmental Protection Agency

Tim Crosby
Slow Money NW

Marcia DeLonge
University of California, Berkeley

Justin Derner
U.S. Department of Agriculture (USDA) Agricultural Research Service (ARS)

Michael Doyle
University of Georgia

Jonathan E. Fielding
County of Los Angeles Department of Public Health

Erin Fitzgerald
Innovation Center for U.S. Dairy

Tracy Fox
Food, Nutrition & Policy Consultants

David Fukuzawa
The Kresge Foundation

Jennifer Greaser
Centers for Disease Control and Prevention

Anne Haddix
Centers for Disease Control and Prevention

James Hammitt
Harvard University

Suzanne Harris
ILSI Research Foundation

Martin Heller
University of Michigan

Sandra Hoffmann
U.S. Department of Agriculture (USDA) Economic Research Service (ERS)

Jill Homer-Stewart
Policy Directions

Helen Jensen
Iowa State University

Steven Kappes
U.S. Department of Agriculture (USDA) Agricultural Research Service (ARS)

Gregory Keoleian
University of Michigan

Steve Kopperud
Policy Directions

Ann Marie Krautheim
National Dairy Council

Robert S. Lawrence
Johns Hopkins University

Mark Lipson
U.S. Department of Agriculture (USDA)

Patty Lovera
Food & Water Watch

Britt Lundgren
Stonyfield Farm

Jayson Lusk
Oklahoma State University

Robert Martin
Johns Hopkins University

Allison Mayor
Environmental Protection Agency

Shalene McNeill
National Cattlemen's Beef Association

Frank Mitloehner
University of California, Davis

Erik Olson
Pew Charitable Trusts

Gifford Pinchot
Bainbridge Graduate Institute

Lance Price
Translational Genomics Research Institute

Mark Rasmussen
Iowa State University

Ricardo Salvador
Union of Concerned Scientists

Beth Sauerhaft
PepsiCo

Jorgen Schlundt
National Food Institute of Denmark

Katherine Smith
American Farmland Trust

Steven Smith
National Institute of Food and Agriculture

Allan Stokes
National Pork Board

Scott Swinton
Michigan State University

Greg Thoma
University of Arkansas

Peter Vardon
FDA Center for Food Safety and Applied Nutrition

Liz Wagstrom
National Pork Producers Council

David Wallinga
Institute for Agriculture and Trade Policy

Aaron Wernham
The Pew Charitable Trusts

Walter Willett
Harvard School of Public Health

Steve Wing
University of North Carolina at Chapel Hill

Hongwei Xin
Iowa State University

Ruihong Zhang
University of California, Davis

IOM/NRC Staff

Allison Berger (IOM)
Geraldine Kennedo (IOM)
Patricia Koshel (NRC)
Janet Mulligan (NRC)
Maria Oria (IOM)
Laura Pillsbury (IOM)
Robin Schoen (NRC)

D

Abbreviations and Acronyms

ARS	Agricultural Research Service
CAFO	concentrated animal feeding operation
CDC	Centers for Disease Control and Prevention
CEAP	Conservation Effects Assessment Project
CPI	Consumer Price Index
EBT	Electronic Benefits Transfer
EPA	Environmental Protection Agency
FDA	Food and Drug Administration
FMPP	Farmers' Market Promotion Program
FOWEL	FOwl WELfare
GHG	greenhouse gas
GRACENET	Greenhouse gas Reduction through Agricultural Carbon Enhancement network
HCSN	Healthy Corner Stores Network
HIA	health impact assessment
IO-LCA	input-output life cycle assessment
IOM	Institute of Medicine
LCA	life cycle assessment

LTAR	Long-Term Agro-Ecosystem Research
LTER	Long-Term Ecological Research Network
MA	Millennium Ecosystem Assessment
NEON	National Ecological Observatory Network
NRC	National Research Council
NSF	National Science Foundation
NSLP	National School Lunch Program
QALY	quality-adjusted life year
SBP	School Breakfast Program
SBP	systolic blood pressure
SNAP	Supplemental Nutrition Assistance Program
SOWEL	SOw WELfare
VOLY	value of a statistical life year
VSL	value of a statistical life
WIC	Special Supplemental Nutrition Program for Women, Infants, and Children
WTP	willingness to pay